TRANSPORTATION INFRASTRUCTURE - ROADS, HIGHWAYS, BRIDGES, AIRPORTS AND MASS TRANSIT

DURABLE COATING SYSTEMS FOR STEEL BRIDGES

TRANSPORTATION INFRASTRUCTURE - ROADS, HIGHWAYS, BRIDGES, AIRPORTS AND MASS TRANSIT

Additional books in this series can be found on Nova's website under the Series tab.

Additional E-books in this series can be found on Nova's website under the E-book tab.

TRANSPORTATION INFRASTRUCTURE - ROADS, HIGHWAYS, BRIDGES, AIRPORTS AND MASS TRANSIT

DURABLE COATING SYSTEMS FOR STEEL BRIDGES

MARGARET S. CRISWELL
EDITOR

nova publishers
New York

For permission to use material from this book please contact us:
Telephone 631-231-7269; Fax 631-231-8175
Web Site: http://www.novapublishers.com

Library of Congress Cataloging-in-Publication Data

ISBN: 978-1-63117-158-1

Published by Nova Science Publishers, Inc. † New York

CONTENTS

Preface

This book focuses on the study and discusses the performance evaluation of one-coat systems for new steel bridges.

Chapter 1 – The Federal Highway Administration 100-Year Coating Study was initiated in August 2009 to identify coating systems that can provide 100 years of virtually maintenance-free service life at comparable costs to the existing coating systems, even in adverse environments. Eight coating systems were selected for the study—three three-coat systems consisting of organic, inorganic, and moisture-cured zinc-based primers; four two-coat systems with various combinations of zinc-based primers and organic top coats; and a single-coat system of calcium sulfonate alkyd. One design innovation feature was the employment of 18- by 18-inch large test panels that contained welding joints and angle attachments using bolts and nuts to closely simulate realistic conditions that are encountered in bridges in the field. All coating systems were evaluated under accelerated laboratory testing and three outdoor exposure conditions, namely natural weathering and natural weathering with salt spray in McLean, VA, and outdoor testing at the Golden Gate Bridge in San Francisco, CA. This report presents results of the performance evaluation of the eight coating systems.

Chapter 2 – In an effort to address cost issues associated with shop application of conventional three-coat systems, the Federal Highway Administration completed a study to investigate the performance of eight one-coat systems and two control coatings for corrosion protection of highway bridges. Based on prior performance, a three-coat system and a two-coat system were selected as the control coating systems. The performance of all coating systems was evaluated under accelerated laboratory and outdoor exposure conditions. Accelerated testing was performed in the laboratory for 6,840 h. Natural weathering exposure was performed in the outdoor environment for 18 months and at a marine exposure site for 24 months. A calcium sulfonate alkyd coating system was found to perform equally in comparison with the three-coat system; however, curing was a major concern. Regression analysis was used to identify correlations between color, gloss, adhesion strength, and coating defects for one-coat systems.

In: Durable Coating Systems for Steel Bridges ISBN: 978-1-63117-158-1
Editor: Margaret S. Criswell

Chapter 1

FEDERAL HIGHWAY ADMINISTRATION 100-YEAR COATING STUDY*

Pradeep Kodumuri and Seung-Kyoung Lee

FOREWORD

The Federal Highway Administration 100-Year Coating Study was initiated in August 2009 to identify coating systems that can provide 100 years of virtually maintenance-free service life at comparable costs to the existing coating systems, even in adverse environments. Eight coating systems were selected for the study—three three-coat systems consisting of organic, inorganic, and moisture-cured zinc-based primers; four two-coat systems with various combinations of zinc-based primers and organic top coats; and a single-coat system of calcium sulfonate alkyd. One design innovation feature was the employment of 18- by 18-inch large test panels that contained welding joints and angle attachments using bolts and nuts to closely simulate realistic conditions that are encountered in bridges in the field. All coating systems were evaluated under accelerated laboratory testing and three outdoor exposure conditions, namely natural weathering and natural weathering with salt spray in McLean, VA, and outdoor testing at the Golden Gate Bridge in

* This is an edited, reformatted and augmented version of The United States Department of Transportation, Federal Highway Administration publication, No. FHWA-HRT-12-044, dated November 2012.

San Francisco, CA. This report presents results of the performance evaluation of the eight coating systems.

Jorge E. Pagán-Ortiz
Director,
Office of Infrastructure Research and Development

SECTION 1. INTRODUCTION

The coating industry switched from lead-based to zinc-based three-coat systems in the 1970s due to health hazards associated with lead in the coatings to protect steel bridges from corrosion. [1] The current state of practice in steel bridge coatings usually involves multi-layer coating typically consisting of a zinc-rich primer over an abrasive blast-cleaned surface and one or two additional coating layers on top of the primer. Both inorganic and organic zinc-rich primers provide galvanic corrosion protection by sacrificing itself to the less electrochemically active steel substrate in the presence of corrosive conditions. The intermediate coat provides a physical barrier to the passage of moisture, oxygen, and other aggressive ions, such as chloride ions, while the top coat protects underlying coating layers against deterioration caused by ultraviolet (UV) radiation and physical damage and enhances the aesthetics of the coating system. It was reported that some three-coat systems with zinc-rich primers can have a service life up to 30 years of protecting steel from corrosion before a major touch-up is required. [2]

Typical costs for conventional coating systems include costs associated with removing mill scale and creating adequate surface condition, coating application, and logistics of moving coated steel bridge members to the field. Such shop operations also require downtime and space during the application of multiple coating layers. The high cost of mill scale removal and surface preparation can be attributed primarily to expensive open-blast cleaning equipment and labor. [2] The cost of typical three-coat systems ranges from under 4 percent to more than 24 percent of the cost of fabricating the steel. [3] After the first shop application, repainting the same bridge after approximately 30 years is very expensive. A practical alternative in reducing these repetitive coating costs is to extend the service life of the shop-applied original steel bridge coatings. [2]

The Federal Highway Administration (FHWA) Coatings and Corrosion Laboratory at the Turner-Fairbank Highway Research Center (TFHRC) initiated an in-house study in August 2009 to identify coating systems that can provide 100 years of virtually maintenance-free service life at comparable costs to the existing coating systems, even in adverse environments. Coating systems in this research study were selected based on past experience and results from previous FHWA studies that evaluated coating systems such as moisture-cured urethanes, waterborne acrylic and epoxy systems, two-coat systems, and one-coat systems. (See references 4–8.) Particularly, previous FHWA in-house studies indicate that the two-coat systems may perform well relative to the well-performing legacy zinc-rich three-coat systems. While critical factors in the field include being cost effective due to a lower number of coats and having reduced application time, two-coat systems have the potential to replace the conventional three-coat systems without sacrificing corrosion protection properties. [9] This report presents research findings from a performance evaluation of eight coating systems based on experimental data from accelerated laboratory testing (ALT) and outdoor exposure testing.

SECTION 2. EXPERIMENTAL PROCEDURES

2.1. Selection of Coating Systems

Table 1 summarizes the eight coating systems employed in this study. Two three-coat systems were used as controls, and the remaining coating systems were test systems composed of a three-coat system, four two-coat systems, and a one-coat system. The acronyms used for all the coating systems are listed in the table. The coating systems will be identified using these acronyms throughout the report.

2.2. Preparation of Test Panels

All of the coating systems were applied onto steel substrates prepared according to the Society for Protective Coatings (SSPC) surface preparation standard number 5 (white metal blast) condition. [10] Subsequently, individual coating systems were applied on the cleaned test panels using an airless spray method by a professional coating laboratory.

Conventional test panels (type I) used for previous FHWA in-house coating studies were 4 by 6 inches. For this study, a new type of test panel (type II) was designed in addition to the conventional test panels to closely simulate detailing of steel bridge members. Each of the type II panels was 18 by 18 inches and contained "V"-shaped and inverted "T"-shaped welding joints, an overlap joint, an angle attachment, and five bolt-nut assemblies, as shown in figure 1 and figure 2. In this report, all type II base plates and individual components were separately spray coated with the primer before panel assembly. After the primer was allowed to dry for 24 h, the components were assembled and coated with an intermediate coat and/or top coat. Introduction of complex geometry to the panel design was employed to understand how coatings applied over the crevices and interfaces created by the attachments, nuts, and bolts of the type II panel perform compared to those applied on the small type I flat panels.

Table 1. Summary of coating systems

System Number	System ID	Coating Type		
		Primer	Intermediate	Top
1	Three-coat (control)	Inorganic zinc-rich epoxy (IOZ)	Epoxy (E)	Aliphatic polyurethane (PU)
2		Zinc-rich epoxy primer (ZE)	E	PU
3	Three-coat	Moisture-cured urethane zinc primer (MCU)	E	Fluorourethane (F)
4	Two-coat	ZE		PU
5		Inorganic zinc primer (Zn)		Polysiloxane (PS)
6		Thermally sprayed zinc primer (TSZ)		linear epoxy (LE)
7		Experimental zinc primer (ZnE)		LE
8	One-coat	High-ratio calcium sulfonate alkyd (HRCSA)		

All type I test panels were coated according to manufacturers' dry film thickness (DFT) recommendations. Each type II panel consisted of three coated areas with the following varying DFT values:

- **Nominal DFT (N)**: Target DFT recommended by the manufacturers.
- **Low DFT (L)**: DFT 20 percent less than the target DFT.
- **High DFT (H)**: DFT 20 percent higher than the target DFT.

Test results from these three DFT areas revealed how much DFT of a particular coating system influenced coating performance. The physical locations of these test areas on the surface of the type II test panel are shown in figure 1.

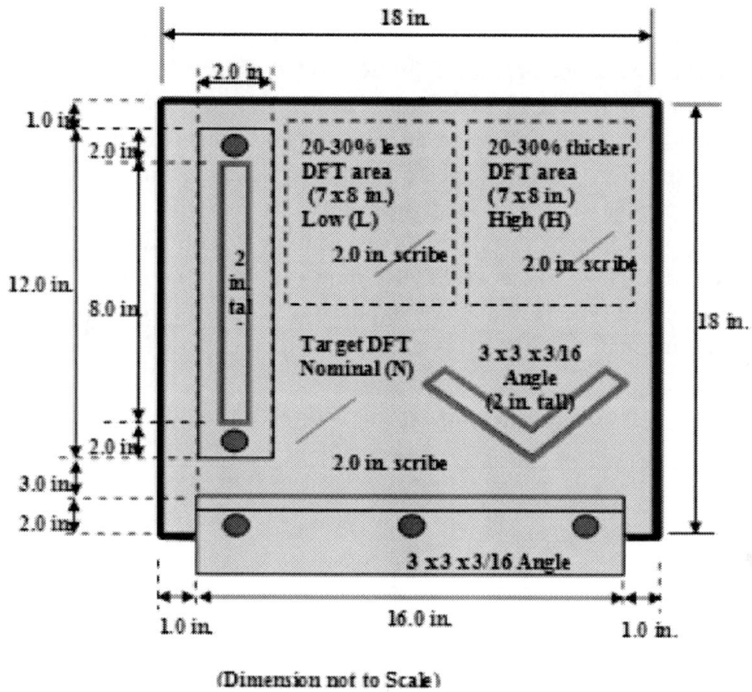

Figure 1. Illustration. Type II test panel.

Figure 2. Photo. Images of type II test panels.

A total of 100 type I and 27 type II test panels were prepared for ALT and outdoor exposure testing. Table 2 and table 3 list the exposure condition and number for type I and II panels, respectively. Outdoor exposure tests were performed in the backyard of TFHRC in McLean, VA, and at the Golden Gate Bridge (GGB) in San Francisco, CA. The TFHRC exposure testing consisted of natural weathering (NW) and natural weathering with daily salt spray (NWS). For type I test panels, each coating system was sprayed on 12 panels, where 5 panels were used for ALT (three scribed and two unscribed), 5 panels were used for NW (one scribed and one unscribed) and NWS (two scribed and one unscribed), and 2 panels were used for physical tests of adhesion strength and Fourier transform infrared spectroscopy (FTIR). In addition, two uncoated type I steel panels were deployed on each of the TFHRC outdoor exposure racks. For type II panels, three panels per coating system were used, one each for the NW and NWS racks at TFHRC and the third one on the GGB rack. Also, an uncoated type II panel was exposed on each of three exposure racks. The overall test matrix of type II panels is provided in table 3.

Table 2. Number of exposure conditions for type I test panels

Group	Number of Coating Systems	ALT	Outdoor Tests		Physical Testing	Total Number of Test Panels
			NW	NWS		
Uncoated steel	N/A	N/A	2	2	N/A	4
Control	2	10	5	5	4	24
Test coating systems	6	30	18	12	12	72
Subtotal	8	40	25	19	16	100

N/A = Not applicable since uncoated steel panels were not tested in ALT and were not used for physical testing.

Table 3. Number of exposure conditions for type II test panels

Group	Number of Coating Systems	Outdoor Tests			Total Number of Test Panels
		NW	NWS	GGB	
Uncoated steel	N/A	1	1	1	3
Control	2	2	2	2	6
Test coating systems	6	6	6	6	18
Subtotal	8	9	9	9	27

N/A = Not applicable since no coating systems were applied on uncoated steel.

After the test panels were delivered to TFHRC, their as-received condition was documented, and six type I panels (three for ALT and three for outdoor testing) for each coating system were scribed following the instructions specified in ASTM D1654-08, "Standard Test Method for Evaluation of Painted or Coated Specimens Subjected to Corrosive Environments." [11] A 2-inchlong scribe was made diagonally on each panel using a mechanical scriber shown in figure 3.

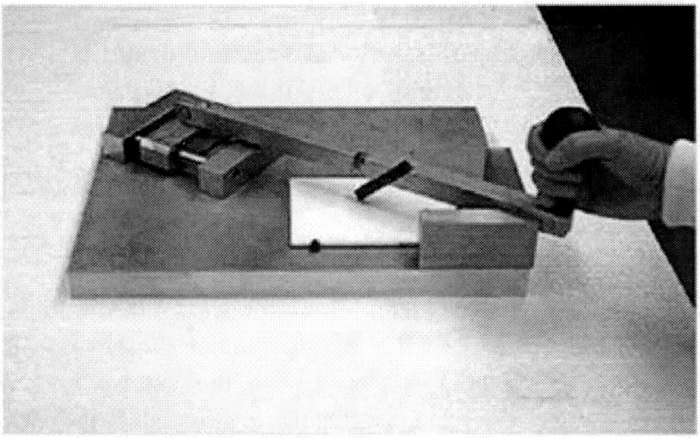

Figure 3. Photo. Scribing tool.

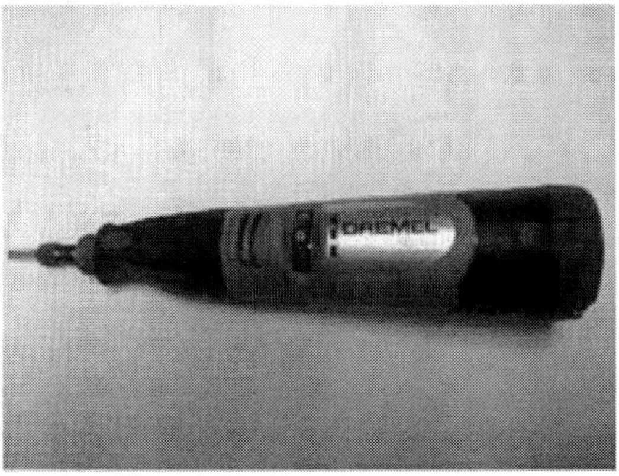

Figure 4. Photo. Dremel® scribing tool.

All test areas of type II test panels were scribed. A mechanical scribing tool used for type I panels was not suited for the large-size test panels. As a result, the test areas were scribed manually using a high-speed Dremel® tool with a scribing bit, as shown in figure 4. C-clamps were used to support a metallic guide along which the scribing was done.

Initial trial scribes were made on dummy test panels to optimize the speed of rotation of the Dremel® bit and applied pressure. The pattern (depth and width) of the scribe made on the dummy test panels was examined using a handheld optical microscope. From this trial, operational parameters were finalized to obtain the ASTM D1654-08 specified width and depth of the scribe, and actual scribes were made on all of the type II panels. [11]

2.3. Test Conditions

ALT

Table 4 summarizes the test conditions for ALT and the total number of cycles. Each ALT cycle was carried out for 360 h, and a total of 20 cycles were planned. Upon completion of every cycle, the panels were examined for their performance. The test panels were also evaluated at the termination of ALT.

Detailed description of each 360-h test cycle is as follows:

- Freeze: 24 h.
- Temperature: -10 °F.
- UV/condensation: 168 h (7 days).
- Test cycle: 4-h UV/4-h condensation cycle.
- UV lamp: UVA-340 nm (wavelength recommended for UV tests).
- UV temperature: 140 °F.
- Condensation temperature: 104 °F.
- Prohesion (cyclic salt-fog, ASTM G85-11, "Standard Practice for Modified Salt Spray (Fog) Testing"): 168 h (7 days). [12]
- Test cycle: 1 h wet/1 h dry.
- Wet cycle: A Harrison mixture of 0.35 weight (wt) percent ammonium sulfate and 0.5 wt percent sodium chloride was used. Fog was introduced at ambient temperature.
- Dry cycle: Air was preheated to 95 °F and was then purged to the test chamber.

Table 4. ALT of type I panels

Item	Freeze Exposure (hours)	UV Condensation Exposure (hours)	Prohesion Exposure (hours)	Total Exposure (hours)
Each cycle	24	168	168	360
Target duration (20 cycles)	480	3,360	3,360	7,200

Figure 5 and figure 6 show a salt fog chamber and a weathering tester, respectively. A 16-h salt-fog accumulation test was conducted before each cyclic salt-fog test to check the atomizing and fog quantity as well as the pH of the collected solution. Typical collection volume of the accumulation test was 0.034–0.068 oz/h, with uniform accumulation in all collection vessels placed at the four corners of the salt-fog chamber and at the center a few feet away from the spray nozzle. Pump speed, throw pressure, and flow rate were adjusted to optimize the collection volume.

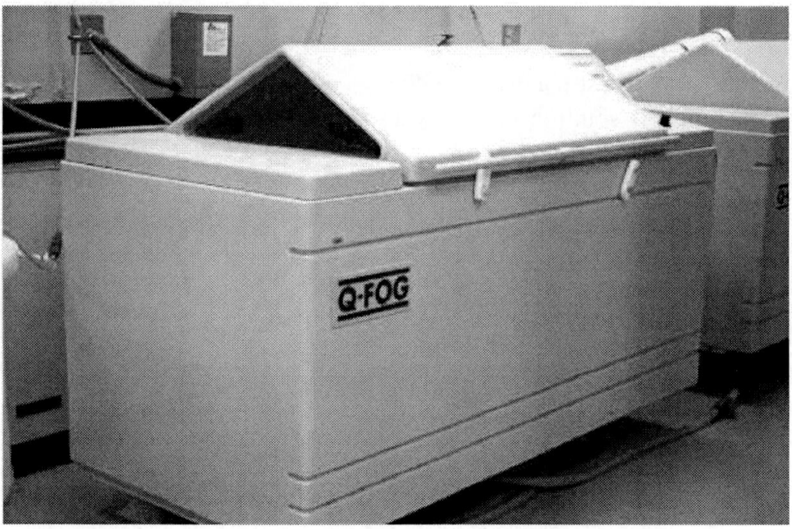

Figure 5. Photo. Salt-fog chamber.

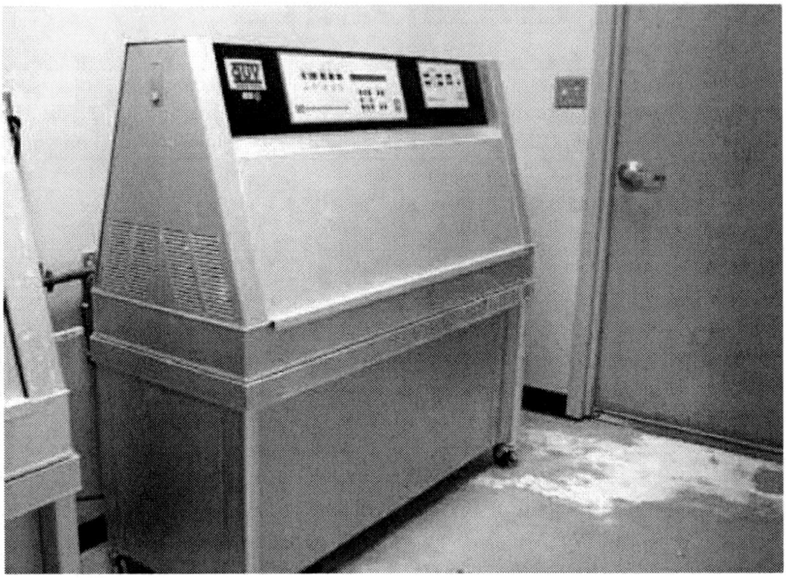

Figure 6. Photo. UV weathering tester.

Type I and II test panels were deployed on two wooden racks inclined at 30 degrees facing south in the backyard at TFHRC in McLean, VA, as shown in figure 7 and figure 8.

Figure 7. Photo. NWS rack at TFHRC.

Figure 8. Photo. NW rack at TFHRC.

In the beginning of the study, NWS test panels received salt spray manually once a day, 5 days per week. Later, an automated daily salt spray system was designed and installed on the NWS rack as shown in figure 9 using the following components:

- 15 wt percent aqueous sodium chloride solution.
- Storage tank for the salt solution (30 gal).
- Electromechanical pump operating in combination with a timer that switches on for a specific time period at a specific time of the day.
- Tubing/piping required for salt solution to flow through to spray onto the deployed panels and a wooden rack.
- Shower heads to spray salt solution onto the deployed panels.

The automated salt spray system provided benefits such as eliminating the need to manually spraying the panels while uniformly spraying all the panels. As a result, daily salt spray could be carried out on weekends and holidays without human intervention. The timer turned the pump on for 15 s of salt spray at 10 a.m. each day. After a week of salt spray, it was clear that the 15 wt percent solution was too severe. Therefore, the salt solution was changed to the Harrison mixture (0.35 wt percent ammonium sulfate and 0.5 wt percent sodium chloride), which was the same solution employed in ALT. Digital pictures of test panels on the racks were recorded every month, and test panels were evaluated for coating performance after 6 months and again at 10 months at the project termination.

Figure 9. Photo. Automated salt spray system for NWS at TFHRC.

Outdoor Exposure Testing at GGB

The outdoor exposure condition at GGB was considered extremely harsh due to severe fog conditions containing airborne chlorides. Figure 10 shows type II panels deployed on top of the south anchorage house near the south abutment. Test panels were scheduled to be evaluated every 6 months for coating performance.

2.4. Coating Characterization Tests and Performance Evaluation Techniques

A series of characterization tests were conducted on the test panels before, during, and after ALT and outdoor exposure tests. Baseline data of DFT, color, gloss, adhesion, and coating defects were collected for each coating system prior to the scheduled tests. Performance of these coating systems was evaluated in terms of development of surface defects and rust creepage during the tests and soon after the completion of the tests. Changes in color, gloss, and adhesion strength were evaluated at the termination of testing. A digital microscope and a digital camera were also employed to document appearance changes of the test panels throughout the study. In addition, electrochemical impedance spectroscopy (EIS) was used to quantify change of coating barrier properties in terms of coating impedance.

Figure 10. Photo. Type II panels deployed at GGB.

DFT

The mean DFT of a coating system was measured in three spots for each type I panel before ALT and outdoor exposure testing using an electronic thickness gauge according to SSPC paint application specification 2, *Measurement of Dry Coating Thickness with Magnetic Gages*. [13] In addition, three spots per each DFT area on the type II panels were obtained to record actual DFTs in each panel.

Gloss

Gloss is defined as the perception of a shiny surface by human eyes. Specular gloss compares the luminous reflectance of a test specimen to that of a standard specimen under the same geometric conditions. [14] Measurements by this test method correlate with visual observations of surface shininess made at roughly the corresponding angles. Measured gloss ratings are obtained by comparing the specular reflectance from the specimen to that from a black glass standard. The measured gloss ratings change as the surface refractive index changes since specular reflectance depends on the surface refractive index of the specimen.

Gloss of all of the coating systems was measured following ASTM D523-08, “Standard Test Method for Specular Gloss.” [15] The 60-degree geometry measurements were conducted on the selected unscribed test panels before and after ALT and outdoor exposure tests. Three gloss readings for each type I panel and six readings for each type II panel were recorded. The reported gloss of each coating system per test condition was the mean of the readings obtained from all unscribed test panels.

Color

The color of the coatings was measured using a colorimeter following ASTM D2244-05, “Standard Practice for Calculation of Color Tolerances and Color Differences from Instrumentally Measured Color Coordinates.” [16] This technique is based on the calculation from instrumentally measured color coordinates based on daylight illumination, color tolerances, and small color differences between opaque coated panels. The Commission Internationale d'Eclairage (International Commission on Illumination or CIE) lab color system (CIE L*, a*, b*) was used for color measurement. L^*, a^*, and b^* represent three coordinates of the three-dimensional lab color space. These parameters are defined based on the high and low values they represent to identify colors as described as follows:

- L* = 0 represents black, while L^* = 100 represents diffuse white.
- Positive values of a* indicate green, while negative values indicate magenta.
- Positive values of b* indicate blue, while negative values indicate yellow.
- The asterisk is used to differentiate the CIE L*, a*, b* system from (L, a, b) parameters of the original Hunter 1948 color space.

Colors were measured for unscribed test panels before and after ALT and outdoor exposure tests. Three color readings were obtained for each type I panel, and six color readings were obtained for each type II panel. Color difference (ΔE) of the test panels before and after the test was calculated using the equation in figure 11.

$$\Delta E = [(\Delta L^*)^2 + (\Delta a^*)^2 + (\Delta b^*)^2]^{1/2}$$

Figure 11. Equation. Measurement of color.

Where:

A L* = L* after test − L* before test. A a* = a* after test − a* before test. A b* = b* after test − b* before test.

The data used in the above equation were the mean of the data obtained from all the unscribed test panels of each coating system.

Adhesion Strength

The adhesion strength of the coating systems was determined using two commercially available pull-off adhesion testers following ASTM D4541-09e1, "Standard Test Method for Pull-Off Strength of Coatings Using Portable Adhesion Testers." [17] A loading fixture, commonly known as a dolly or stub, is affixed to the panel surface by an adhesive. A load provided by the adhesion tester is increasingly applied to the dolly until it is pulled off. The force required to pull the dolly off yields the tensile strength in pounds per square inch (psi). Failure will occur along the weakest plane(s) within the testing system comprised of the dolly, adhesive, individual layers of the coating system, and substrate.

The surface of coated test panels and the base of the dollies were cleaned with detergent water and were lightly roughened with an abrasive pad. The dollies were glued on the test panel surface using a high-strength epoxy adhesive. The cut through the coating around the edge of the dolly was made using a drill press after the complete curing of the adhesive, as seen in figure 12. The initial and final adhesion strengths of coating systems were measured by hydraulic method. Figure 13 shows the hydraulic adhesion tester used in this study. For each coating system, three pull-off adhesion tests were performed on two unscribed virgin type I panels and on each of the tested type I panels. No adhesion strength tests were performed on type II panels.

For every panel, the average adhesion strength of three locations was calculated. If the coefficient of variance (CV) of each test panel was more than 20 percent, the test panel and adhesion failure mode were carefully examined to see if the variation was caused by test operation. Repeated tests were performed for quality assurance. If more than 50 percent of a glue failure occurred, the test was also repeated. The reported adhesion strength for each coating system was the mean of the data obtained from tests conducted on all test panels of the coating system. The remaining DFT at the pull-off spots was measured and recorded. The adhesion failure mode of every spot was also documented using digital photographs.

Figure 12. Photo. Drill press to score a test area around a dolly.

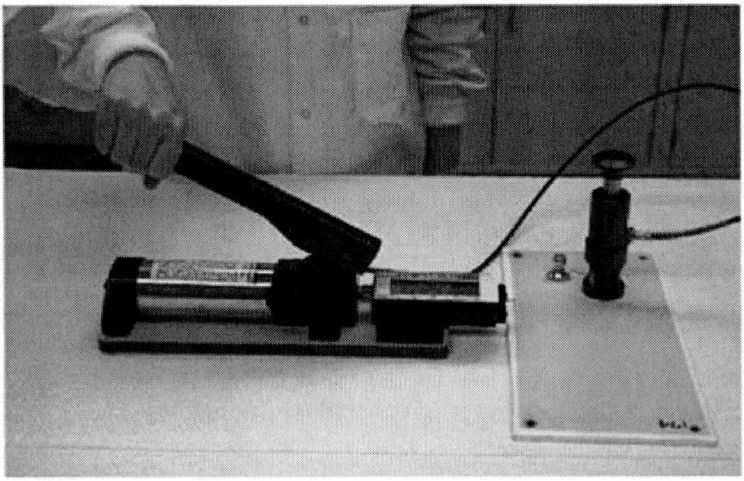

Figure 13. Photo. Hydraulic adhesion tester.

Detection of Coating Defects

The coating defects were identified according to ASTM G62-07, "Standard Test Methods for Holiday Detection in Pipeline Coatings (Method A)." [18] This technique utilizes a low voltage holiday detector to determine the presence of electrically conductive coating defects including holidays (invisible defects with naked eyes) and pinholes, voids, mechanical coating damage, or metal particles protruding through the coating. The reported number of defects after each test cycle was the cumulative number of defects. In addition to using the holiday detector, test panels were visually examined for blisters and rust spots per ASTM D714-02, "Standard Test Method for Evaluating Degree of Blistering of Paints" and ASTM D610-01, "Standard Test Method for Evaluating Degree of Rusting on Painted Steel Surfaces." [19,20] The reference standards were employed to grade the rust pits and surface blisters on the panels.

Rust Creepage Measurement

The rust creepage at the scribe was measured following ASTM D7087-05a, "Standard Test Method for An Imaging Technique to Measure Rust Creepage at Scribe on Coated Test Panels Subjected to Corrosive Environments." [21] The rust creepage area from the scribe line on the coating panel was traced using a thin marker and a transparent plastic sheet. The tracing image was scanned and analyzed using imaging software to obtain the creepage areas and the creepage distances. Two traces for each test panel were obtained, and the mean creepage distance was reported as the nominal creepage at the time of measurement for the coating system.

Digital Microscopic Examination

When unusual surface failures were detected by the holiday detector, such panels were examined using a stereomicroscope or a high-power digital microscope. The surface conditions were documented via microphotographs.

Digital Photography

Every test panel was photographed to document surface conditions before initiating the tests as well as after each test cycle for both ALT and outdoor exposure tests.

EIS

The impedance of the coating systems was measured by EIS using an electrochemical instrument equipped with a potentiostat. This technique involves applying a small amplitude alternating current signal into a body of material over a wide range of frequencies and measuring the responding current and its phase angle shift. The output from the EIS instrument is an impedance spectrum of the material, typically ranging from 100 kHz to 0.001 Hz. EIS data are analyzed by the equivalent circuit modeling technique, which can produce appropriate models to evaluate the coating deterioration process and the mechanism of corrosion occurring at the interface between the substrate and the coating. Analysis of the EIS data is not included in this report, and the analysis results will be presented in a separate report in the future.

SECTION 3. TEST RESULTS AND DISSCUSION

Initial coating characterization results and test results from ALT and outdoor exposure tests pertaining to color, gloss, DFT, pull-off adhesion, surface appearance, surface defects, and rust creepage are presented in this Section with some supplementary digital images.

3.1. Initial Characterization of Coating Systems

DFT

Table 5 and figure 14 show mean DFT, standard deviation, and CV of the coating systems measured on type I panels. DFT varied significantly from 7.5 to 17.8 mil depending on the coating system.

Table 5. DFT data for type I panels

Coating System	Number of Coats	Mean (mil)	Standard Deviation (mil)	CV (Percent)
IOZ/E/PU	3	13.03	0.66	5.07
ZE/E/PU	3	14.31	0.84	5.87
MCU/E/F	3	13.31	0.67	5.03
ZE/PU	2	7.75	0.56	7.23
Zn/PS	2	7.5	0.46	6.13
TSZ/LE	2	15.97	1.31	8.2
ZnE/LE	2	17.87	1.26	7.05
HRCSA	1	8.3	0.53	6.39

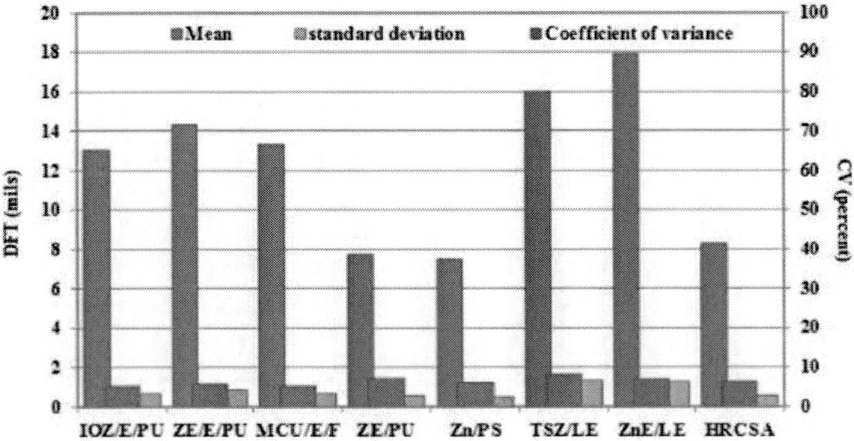

Figure 14. Graph. DFT data (type I panels).

None of the mean DFTs was less than 7 mil, and two of the two-coat systems (ZE/PU and Zn/PS) were between 7 and 8 mil. All three-coat systems (IOZ/E/PU, ZE/E/PU, and MCU/E/F) had mean DFTs between 13 and 15 mil. The remaining two-coat systems (TSZ/LE and ZnE/LE) had the highest DFT values (16–18 mil) and the largest variation of DFT data in terms of CV and standard deviation. The one-coat system (HRCSA) had a DFT of 8.3 mil, which was slightly thicker than two of two-coat systems, ZE/PU and Zn/PS.

Gloss

Figure 15 shows the initial mean gloss readings of type I and type II panels. The eight coating systems exhibited a wide range of initial gloss, with HRCSA having the lowest gloss reading of 9.5 and MCU/E/F, Zn/PS, TSZ/LE, and ZnE/LE having the high gloss readings greater than 70. The other coating systems, IOZ/E/PU, ZE/E/PU, and ZE/PU, possessed intermediate gloss. In most cases, the coating systems exhibited different gloss readings depending on the type of panel (i.e., type I versus type II). This observation suggests that absolute gloss readings may not be useful due to inherent variation of the measurement technique.

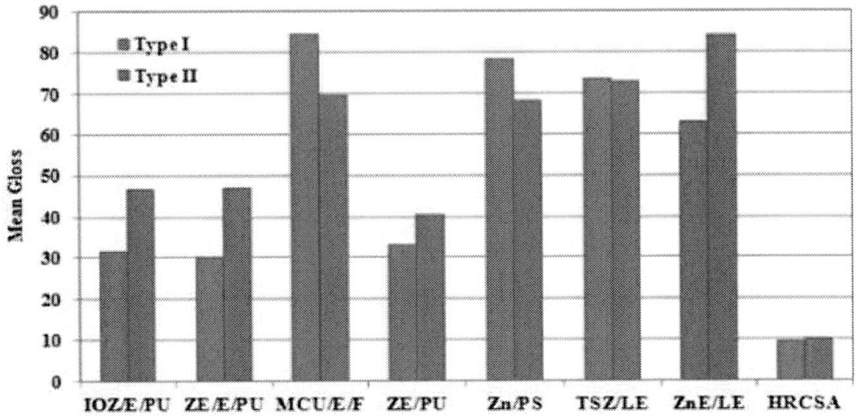

Figure 15. Graph. Initial mean gloss.

Color

Initial mean color readings are summarized in table 6. To calculate change of color (ΔE), color readings made at different times were necessary, as shown in the equation in figure 11.

Table 6. Initial mean color readings

Coating System	Number of Coats	Type 1			Type II		
		L*	a*	b*	L*	a*	b*
IOZ/E/PU	3	68.83	-1.64	6.20	68.70	-1.68	6.47
ZE/E/PU	3	68.84	-1.64	6.18	68.82	-1.58	6.46
MCU/E/F	3	92.66	-0.79	1.23	92.23	-0.84	1.50
ZE/PU	2	68.86	-1.64	6.13	68.72	-1.67	6.46
Zn/PS	2	96.55	-1.35	4.15	94.20	-1.40	3.88
TSZ/LE	2	47.56	-2.72	-0.64	47.08	-2.82	0.46
ZnE/LE	2	48.1	-2.80	-0.26	47.10	-2.81	0.16
HRCSA	1	61.49	-0.69	1.71	61.30	-0.75	1.84

Adhesion Strength

Initial mean adhesion values are shown in table 7 and figure 16. Two of the two-coat systems (ZnE/LE and TSZ/LE) exhibited the highest baseline adhesion strengths of 2,110 and 1,834 psi (CV = 21 and 13 percent), respectively, while the one-coat system (HRCSA) demonstrated the lowest adhesion strength of 259 psi (CV = 45 percent). The poor adhesion strength of HRCSA was observed in a previous FHWA study. [22] The two control coating systems (IOZ/E/PU and ZE/E/PU) had the second and third lowest adhesion strengths of 1,119 and 1,173 psi (CV = 36 and 30 percent), respectively. All remaining three-coat and two-coat systems showed moderate adhesion strengths between 1,200 and 1,400 psi (CV = 17 to 26 percent).

Table 7. Initial pull-off adhesion strength of type I panels

Coating System	Number of Coats	Average Adhesion Strength (psi)	CV (Percent)	Standard Deviation (psi)
IOZ/E/PU	3	1,119	36	400
ZE/E/PU	3	1,173	30	354
MCU/E/F	3	1,371	26	356
ZE/PU	2	1,314	19	250
Zn/PS	2	1,259	17	213
TSZ/LE	2	1,834	13	242
ZnE/LE	2	2,110	21	433
HRCSA	1	259	45	115

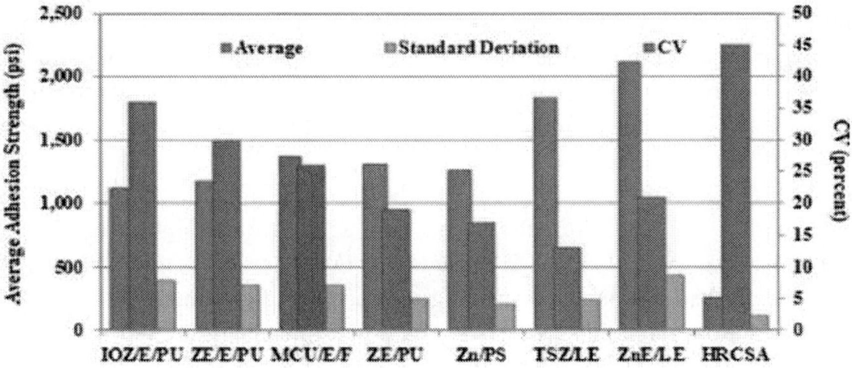

Figure 16. Graph. Initial pull-off adhesion strength data (type I panels).

3.2. Performance during ALT and Outdoor Exposure Testing

Performance of the coating systems in ALT, NW, NWS, and GGB was evaluated by the following parameters:

- Gloss reduction.
- Change of color.

- Change of adhesion strength.
- Development of surface defects.
- Growth of rust creepage at the scribe.

Due to premature coating failures of two of the two-coat systems (TSZ/LE and ZnE/LE) in ALT, the study was terminated after 10 cycles of ALT (3,600 h), 10 months of NW and NWS, and 6 months of GGB exposure. Unexpected early termination of the study led to no data collection from type II panels used for 6 months at GGB. Therefore, most outdoor performance is discussed with the NW and NWS data only.

Gloss Reduction

At the termination of the study, gloss data of each coating system were obtained from type I and type II panels exposed to ALT, NW, and NWS environments. Then, gloss reductions of the coating systems after 3,600 h of ALT and 10 months of NW and NWS testing were calculated with respect to their initial readings. Table 8 and figure 17 present the gloss reduction data. The largest gloss reduction was exhibited by two two-coat systems (TSZ/LE and ZnE/LE) and the one-coat system, HRCSA, as shown in figure 17. These coating systems experienced the most dramatic changes from the highest initial gloss readings (two-coat systems) and the lowest gloss reading (one-coat system). It should be noted that the two-coat systems failed prematurely, and both had the same linear epoxy top coat. Conversely, two other coating systems (MCU/E/F and Zn/PS), which also exhibited high initial gloss readings (see figure 15), exhibited the least amount of changes after testing. The three-coat controls (IOZ/E/PU and ZE/E/PU) and the two-coat system ZE/PU exhibited intermediate initial gloss readings and remained at less than 30 percent gloss reduction after tests. The type II panels in NW tended to show the largest gloss reduction among coating systems, with little to moderate overall gloss reductions.

Color Changes

Table 9 and figure 18 show mean color change data. HRCSA exhibited the largest color change, followed by two two-coat systems (TSZ/LE and ZnE/LE). These coating systems exhibited the largest gloss reductions. The others exhibited less than 2 percent color changes after testing. Because of scattered data among different test conditions, effects of salt spray and type of test panel on color change were inconclusive, except that most coating systems exhibited the least amount of color changes in ALT.

Table 8. Mean gloss reduction (percent)

Coating System	Number of Coats	Type 1			Type II		
		ALT	NW	NWS	NW	NWS	GGB
IOZ/E/PU	3	12.96	9.47	27.6	27.96	15.15	No data available
ZE/E/PU	3	17.76	14.51	8.89	44.93	20.87	
MCU/E/F	3	0.82	3.65	2.13	30.4	3.17	
ZE/PU	2	14.7	10.07	9.83	24.62	22.28	
Zn/PS	2	-6.88	1.42	-1.61	10.1	3.3	
TSZ/LE	2	94.73	92.63	92.47	95.54	94.97	
ZnE/LE	2	75.6	92.57	96.2	95.39	96.39	
HRCSA	1	24.41	66.9	50.57	68.91	74.81	

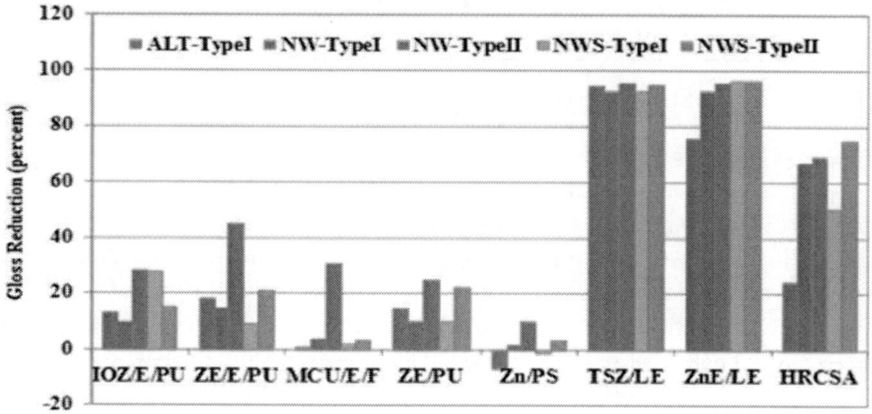

Figure 17. Graph. Mean gloss reduction.

Table 9. Mean color changes (percent)

Coating System	Number of Coats	Type 1			Type II	
		ALT	NW	NWS	NW	NWS
IOZ/E/PU	3	0.08	1.34	0.60	0.67	6.43
ZE/E/PU	3	0.32	1.03	0.46	0.63	0.40
MCU/E/F	3	0.12	1.07	0.44	0.27	1.01
ZE/PU	2	0.16	1.16	0.37	0.35	0.71
Zn/PS	2	1.29	0.71	1.25	1.40	0.93
TSZ/LE	2	3.93	3.34	4.60	4.35	3.52
ZnE/LE	2	2.91	2.64	4.04	4.24	3.10
HRCSA	1	2.66	5.14	7.15	7.37	11.96

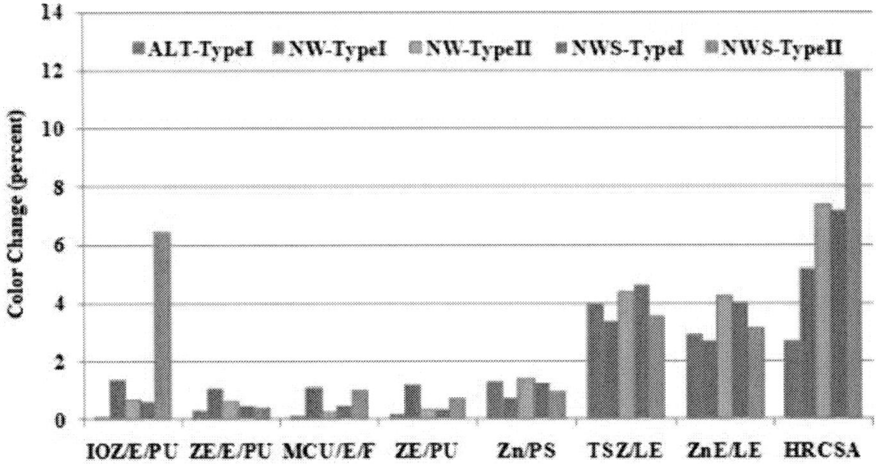

Figure 18. Graph. Mean color changes.

Changes in Adhesion Strength

Adhesion strength changes in type I panels after exposure in ALT, NW, and NWS are summarized in table 10, and the corresponding bar graph is shown in figure 19. All coating systems, except HRCSA and MCU/E/F, showed varying degrees of adhesion strength reduction at the end of testing in every test condition. No clear trend was observed between reduction of adhesion strength and test conditions. It is unclear at this time why HRCSA, which exhibited the weakest initial adhesion strength, had the largest strength gain during testing. However, it may be reasoned that because HRCSA had the weakest coating before and after testing, relatively small increases in post-test absolute adhesion strength resulted in significant increases in terms of percentage. Such a trend was more pronounced for NW and NWS panels than ALT panels. Adhesion strength data for TSZ/LE could not be obtained at the termination of ALT because the coating surface was so blistered that the adhesion dollies did not stick to the surface of the panels. Because ZnE/LE panels had peeling top coats at the end of testing, dollies were carefully placed in areas with intact top coats. ZnE/LE panels had the largest adhesion strength reduction. The two worst performing two-coat systems (TSZ/LE and ZnE/LE) in gloss and color also suffered from the largest adhesion reductions. The least amount of adhesion strength reduction was seen in both controls (IOZ/E/PU and ZE/E/PU) as well as ZE/PU. The other two-coat system, Zn/PS, had a moderate adhesion strength loss.

Table 10. Mean adhesion strength changes of type I panels

Coating System	Number of Coats	Initial Adhesion Strength (psi)	Final Adhesion Strength in Exposure Condition (psi)			Percent Change		
			ALT	NW	NWS	ALT	NW	NWS
IOZ/E/PU	3	1,119	818	1,073	1,108	-27	-4	-1
ZE/E/PU	3	1,173	1,135	1,124	1,365	-3	-4	-16
MCU/E/F	3	1,371	1,423	1,561	1,618	4	14	18
ZE/PU	2	1,314	1,176	1,295	1,258	-10	-1	-4
Zn/PS	2	1,259	976	1,036	1,024	-22	-18	-19
TSZ/LE	2	1,834		1,150	911		-37	-50
ZnE/LE	2	2,110	651	1,321	1,207	-69	-37	-43
HRCSA	1	259	312	461	506	21	78	96

Note: Blank cells indicate that final adhesion strength and percent change could not be obtained for TSZ/LE at the termination of ALT.

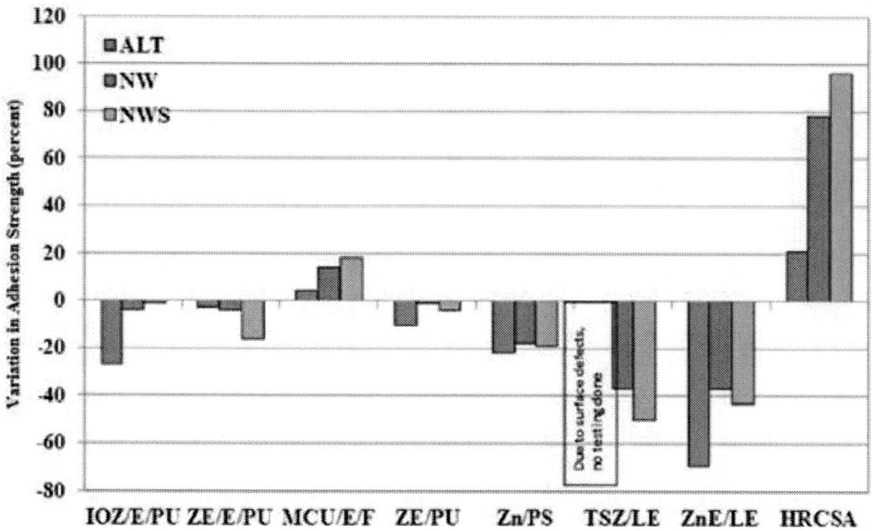

Figure 19. Graph. Mean adhesion strength changes of type I panels.

Development of Surface Defects

ALT

The total number of cumulative surface defects that developed during ALT is listed in table 11, and the corresponding line-scatter graph is shown in figure 20. The exact number of defects could not be counted when an excessive amount was detected. As a result, an arbitrary number of 100 was entered in the data sheet. Excessive defects were observed on the surface of two two-coat system (TSZ/LE and ZnE/LE) test panels. Even though initial assessment of these coating systems showed no defects, including invisible defects (holidays), on the surface, a sudden increase in the number of defects during ALT led to severe surface deterioration as indicated by blistering, rusting, and/or cracking, which is discussed in the next section.

In the case of TSZ/LE, one of the panels developed four defects after 1,080 h of testing, and the number of defects increased excessively after 1,440 and 1,800 h of testing. These surface defects were then followed by excessive blistering and cracking of the surface. For ZnE/LE, no defects were detected until 2,880 h of testing. Progressive changes were observed leading to surface microcracks, which transformed to macrocracks. As shown in table 11, MCU/E/F did not develop any defects, while other coating systems only developed minimal coating defects upon termination of ALT at 3,600 h.

Table 11. Cumulative number of surface defects developed during ALT

Time (hours)	IOZ/E/PU	ZE/E/PU	MCU/E/F	ZE/PU	Zn/PS	TSZ/LE	ZnE/LE	HRCSA
0	0	0	0	0	0	0	0	0
360	0	1	0	0	0	1	0	1
720	0	1	0	0	0	1	0	1
1,080	0	1	0	0	0	4	0	1
1,440	0	1	0	0	1	100	0	1
1,800	0	1	0	1	1	100	0	1
2,160	0	1	0	1	1	100	0	1
2,520	0	1	0	1	1	100	0	1
2,880	2	1	0	1	1	100	2	1
3,240	2	2	0	3	2	100	100	1
3,600	2	2	0	3	2	100	100	1

NW and NWS

Most of the type I panels did not develop any defects during outdoor exposure in NW and NWS. Exceptions were two of the two-coat systems, TSZ/LE and ZnE/LE, which performed poorly in other tests. TSZ/LE exhibited countless defects on every NW and NWS type I panel. For ZnE/LE, five defects on NW panels and seven defects on NWS panels were developed after 10 months of outdoor testing. All type I and type II test panels employed in NW and NWS developed many defects along the panel edges. Type II panels also had coating defects in areas such as nuts, bolts, underside of the T-attachment, and the wide-angle attachment due to less than ideal coating application conditions. As expected, these defects became rust spots. This finding confirms that it is difficult to avoid initial coating defects from coating applications on field bridge structures due to complex shapes of structural elements. These imperfect sites are prone to advanced coating failures and subsequent steel corrosion in service environments.

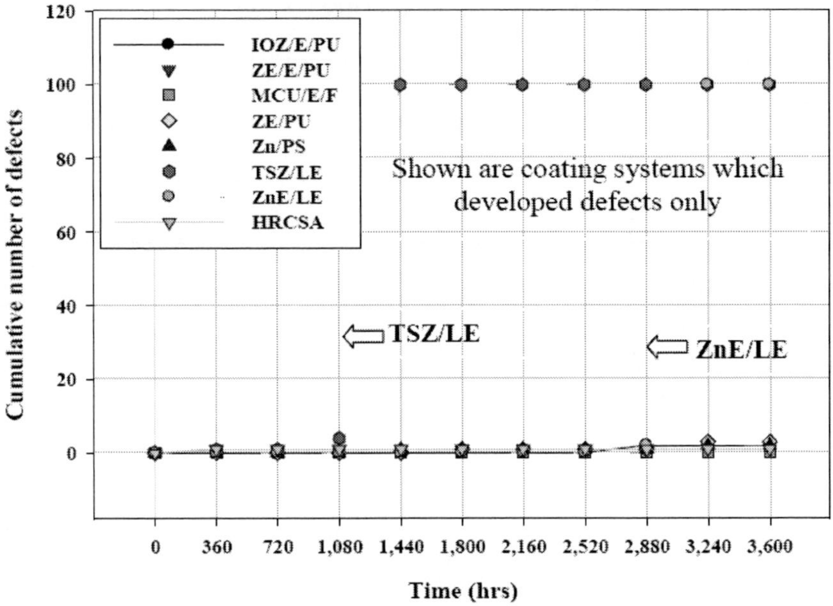

Figure 20. Graph. Cumulative number of surface defects developed on type I panels during ALT.

Table 12 lists coating defect development for type II panels. For the sake of simplicity in this report, only DFT-measured flat areas on all panels were studied for defect development. Zn/PS initially contained numerous coating defects in the low DFT areas, and none of the high and normal DFT areas developed additional defects during outdoor exposure testing. Although TSZ/LE started with a moderate number of coating defects in the low DFT and six in the normal DFT areas, they ended up with numerous defects on every panel regardless of DFT and exposure condition. Also, the low DFT area of the ZnE/LE coating system in NWS developed greater than 70 percent surface deterioration accompanied by surface cracking and coating failure. HRCSA started with 3 initial defects and developed 15 defects in the low DFT and 1 defect in the normal DFT areas in both NW and NWS. The 15 defects in the NW and NWS environments appeared to be related to mechanical damage through the soft coating. The other coating systems did not develop additional coating defects during the 10 months of exposure testing. Coating defect data were not available for the panels at the GGB exposure site.

Table 12. Number of surface defects developed on type II panels during NW and NWS exposure testing

Coating System	Exposure	Initial			After 10 months of Exposure Testing		
		L	H	N	L	H	N
IOZ/E/PU	NWS	0	0	0	0	0	0
	NW	0	0	0	0	0	0
ZE/E/PU	NWS	2	0	0	2	0	0
	NW	0	0	0	0	0	0
MCU/E/F	NWS	0	0	0	0	0	0
	NW	0	0	0	0	0	0
ZE/PU	NWS	0	0	0	0	0	0
	NW	0	0	0	0	0	0
Zn/PS	NWS	Numerous defects	0	0	Numerous defects	0	0
	NW	Numerous defects	0	0	Numerous defects	0	0
TSZ/LE	NWS	> 50	0	6	Numerous defects	Numerous defects	Numerous defects
	NW	> 25	0	0	Numerous defects	Numerous defects	Numerous defects
ZnE/LE	NWS	0	0	0	> 70 percent failure	0	0
	NW	0	0	0	0	0	0
HRCSA	NWS	0	0	1	6	0	1
	NW	0	0	2	8	0	3

Rust Creepage

ALT

Table 13 and figure 21 show rust creepage data measured during 3,600 h of ALT. Because all of the TSZ/LE panels started to develop serious surface deterioration after only 1,080 h of ALT, rust creepage data from this coating system were not measured.

Based on the mean creepage values at the end of 3,600 h of ALT, the coating systems were ranked in the following order from highest to lowest rust creepage:

MCU/E/F >> ZE/E/PU ≈ Zn/PS ≈ ZE/PU ≈ ZnE/LE > IOZ/E/PU > HRCSA

It should be noted that ZE/E/PU creepage data showed a steadily decreasing trend during the last three cycles, which was caused by measurement errors. Based on a rapid creepage growth of ZnE/LE observed from the sixth cycle (2,520 h), it can be projected that the coating system would have shown a very high rust creepage if ALT had continued for more cycles. As observed in previous a FHWA one-coat study, HRCSA had the best performance for resistance to rust creepage. [22]

Table 13. Rust creepage growth during ALT (mm)

Time (hours)	IOZ/E/PU	ZE/E/PU	MCU/E/F	ZE/PU	Zn/PS	ZnE/LE	HRCSA
0	0	0	0	0	0	0	0
360	0	0	0	0.09	0	0	0.18
720	0	0.66	0	0.29	0	0	0.25
1,080	0	0.59	0.82	0.43	0.22	0	0.24
1,440	0	0.88	0.88	0.45	0.37	0	0.25
1,800	0.27	1.15	1.08	0.59	0.49	0	0.29
2,160	0.29	1.42	2.28	0.97	0.60	0	0.29
2,520	0.64	1.25	2.05	0.98	0.93	0.86	0.31
2,880	0.73	1.42	3.14	1.11	1.03	1.02	0.32
3,240	0.74	1.36	3.43	1.23	1.07	1.07	0.32
3,600	0.77	1.30	3.80	1.20	1.24	Panel removed	0.33

1 inch = 25.4 mm.

NW and NWS

None of the type I panels developed any rust creepage during NW and NWS outdoor exposure testing except for ZnE/LE, which had a rust creepage of 0.16 inches at the end of NWS exposure. After 10 months of exposure of type II panels in NW and NWS testing and 6 months of exposure at the GGB, only the ZnE/LE coating system exhibited recognizable rust creepage, as shown in table 14 and figure 22. High DFT areas of type II panels with ZnE/LE in NWS showed extremely high creepage of around 0.51 inches. Similarly, the nominal DFT area of the same panel showed creepage of about 0.23 inches. The rest of the type II coating systems did not show any rust creepage. It is interesting to observe that ZnE/LE, which was mediocre in ALT, showed the worst rust creepage performance in NW and NWS, whereas MCU/E/F, which performed the worst in ALT, performed well in NW and NWS without rust creepage.

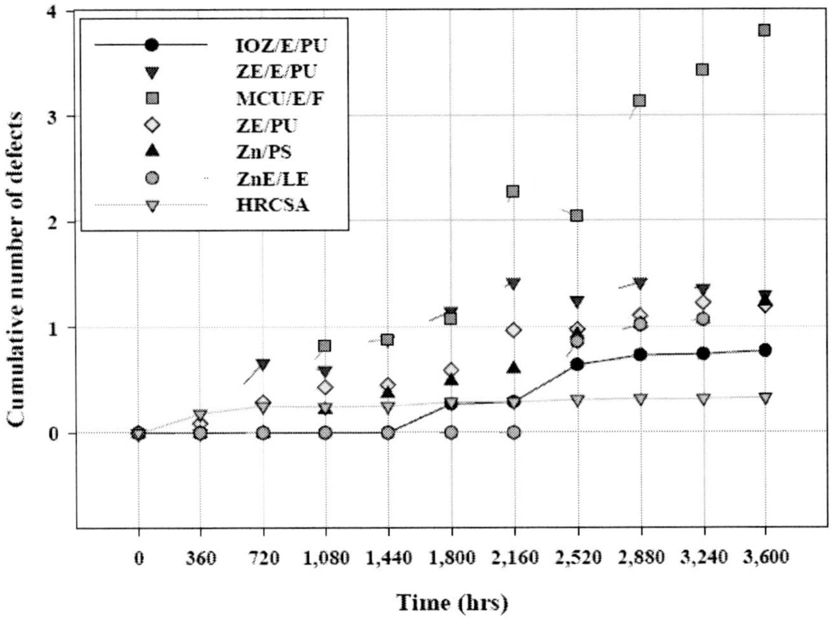

Figure 21. Graph. Rust creepage growth with time during ALT.

Table 14. Rust creepage growth of ZnE/LE during outdoor exposure

Coating System	Exposure	DFT Area	Rust Creepage Area (inches^2)
ZnE/LE (type I)	NWS	N	0.52
ZnE/LE (type II)	NW	N	0.30
	NWS	H	1.59
		L	0.46
		N	0.65
	GGB	H	0.30

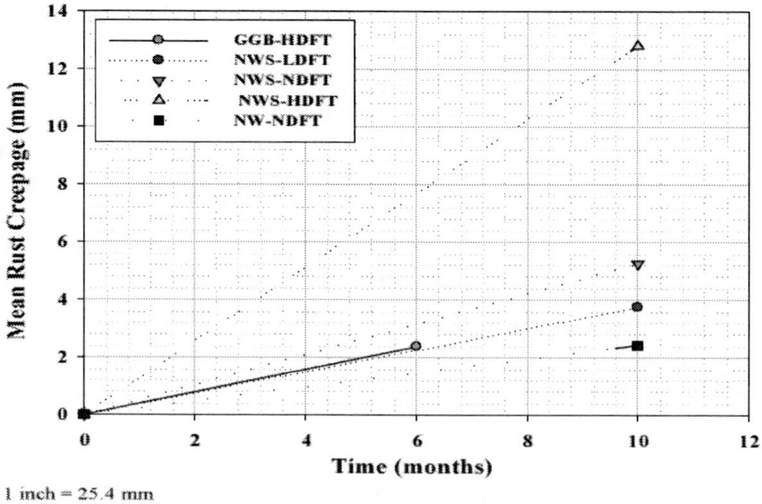

Figure 22. Graph. Rust creepage growth of ZnE/LE (type II panels).

Physical Condition of Representative Test Panels

Figure 23 through figure 30 show typical digital photographs of all coating systems for type I panels during ALT at 0, 1,440, 2,520, and 3,600 h. These figures also show representative photographs of the type I panels in NW and NWS before and at the termination of testing at 10 months.

ALT

The two controls, IOZ/E/PU and ZE/E/PU (see figure 23 and figure 24), had the best surface retention properties in all exposure conditions despite the fact that ZE/E/PU exhibited the second highest rust creepage. The one-coat system, HRCSA, also retained good surface characteristics with the least rust creepage and minimal defect development despite the fact that it exhibited the third largest gloss reduction and the highest color change (see figure 30). As observed in a previous FHWA one-coat study, HRCSA remained very soft throughout the testing. [22] The third three-coat system, MCU/E/F did not develop defects, but it had the highest creepage at the termination of ALT (see figure 25). Two of the two-coat systems, ZE/PU and Zn/PS, had moderate defect development and rust creepage growth (see figure 26 and figure 27). The remaining two-coat systems, TSZ/LE and ZnE/LE, had the worst surface deterioration over time in ALT as indicated by the development of defects, surface blistering, coating peel-off, and/or surface cracking (see figure 28 and figure 29). The visual observation of surface changes was followed up with digital microscopy of the test panel surfaces. Digital microscopy of TSZ/LE showed the various phases of progressive deterioration of the coating system at 0, 1,800, and 3,600 h (see figure 31). At 1,800 h, the surface of the coating started to develop blisters, which progressively grew to form a completely blistered coating surface at the termination of ALT, as shown in figure 31. Certain areas still had blisters intact, while some areas had half-peeled or detached coating from the surface. White residual zinc oxide formed on the surface where the coating had peeled off.

Digital microscopy of ZnE/LE during ALT yielded unexpected critical observations. The surface of the coating system showed microcracks all over the surface of a test panel, as shown in figure 32. The central points at which these microcracks originated were a few mils in size, and the cracks themselves were a few hundred mils in size. The microcracks developed into surface macrocracks at the termination of ALT (see figure 32). However, these test panels did not show any surface defects indicated by closed circuit shorting (beeping) during holiday testing, indicating that the cracks did not develop through the coating thickness and might have formed superficially near the surface. Comparison of digital microscopy images before and after exposure indicated that these crack-originating locations were present on the panel surface before testing. However, cracks that propagated from these locations appeared only after ALT. The density of these cracks did not seem to grow as ALT progressed.

NW and NWS

Figure 23 through figure 30 also show type I panels exposed to NW and NWS. In general, their physical conditions appeared to be better than their counterparts in ALT. TSZ/LE exposed to NWS formed distinctive white zinc oxide within the scribe. Additionally, ZnE/LE and HRCSA under NWS exposure exhibited heavy rust stains in some areas. Interestingly, none of the TSZ/LE and ZnE/LE type I panels, shown in figure 28 and figure 29, respectively, suffered from the same surface deteriorations of blistering (see figure 31) and cracking (see figure 32) observed on their ALT counterparts. Such a finding suggests that either different deterioration mechanisms worked in ALT and NW/NWS or that 10 months in outdoor exposure testing was not sufficient to introduce significant deterioration.

Figure 33 through figure 40 show digital photographs of all coating systems on type II panels before exposure and at 6 months of outdoor exposure testing at GGB. These figures also show conditions of panels before and after exposure in NW and NWS for 10 months. Although each type II panel consisted of low, high, and nominal DFT areas, rust was observed predominantly at hard-to-reach areas where coating was non-uniformly applied and/or where there were coating defects (bare areas). These areas included nuts, bolts, underside of the T-attachment, and the wide-angle attachment. TSZ/LE showed formation of white zinc oxide within the scribe on all DFT areas as well as along the edges of nuts and bolts and underneath the attachments (see figure 38). The ZnE/LE panel exhibited excessive rusting on all DFT areas and along the edges of nuts and bolts and underneath the attachments in NWS (see figure 39). The same panel also showed rust buildup within the V-notch section where salt water could be collected. This form of deterioration reveals that ZnE/LE was particularly venerable to prolonged exposure to salt water. The low DFT area of this coating system developed cracks all over the surface, which was similar to what was observed toward the end of ALT (see figure 32). The two controls did not show any signs of rusting within the three DFT areas (see figure 33 and figure 34). As observed in ALT, the next best performing coating system in comparison to the two three-coat controls was HRCSA, as indicated by negligible rusting and rust creepage (see figure 40).

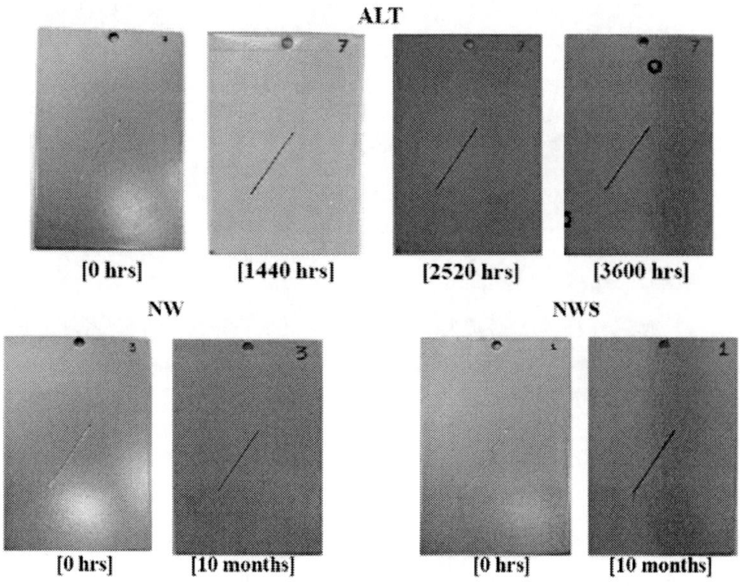

Figure 23. Photo. Progressive changes of IOZ/E/PU—type I in ALT, NW, and NWS.

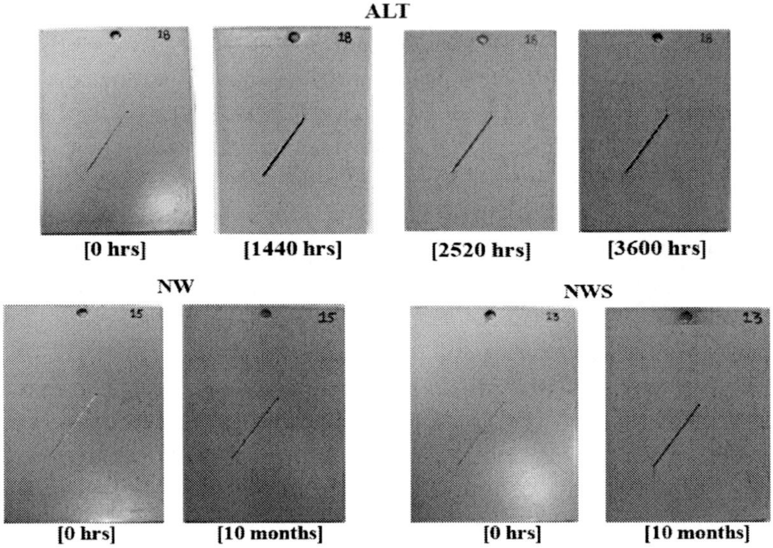

Figure 24. Photo. Progressive changes of ZE/E/PU—type I in ALT, NW, and NWS.

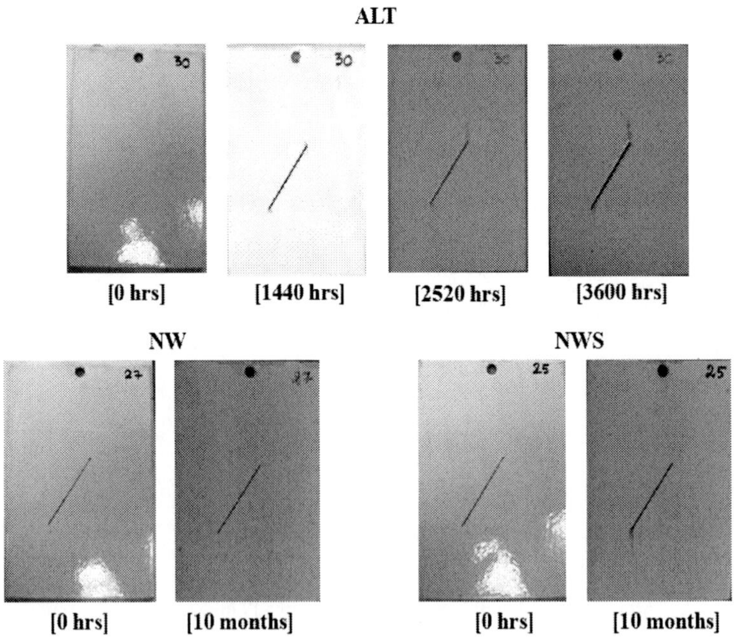

Figure 25. Photo. Progressive changes of MCU/E/F—type I in ALT, NW, and NWS.

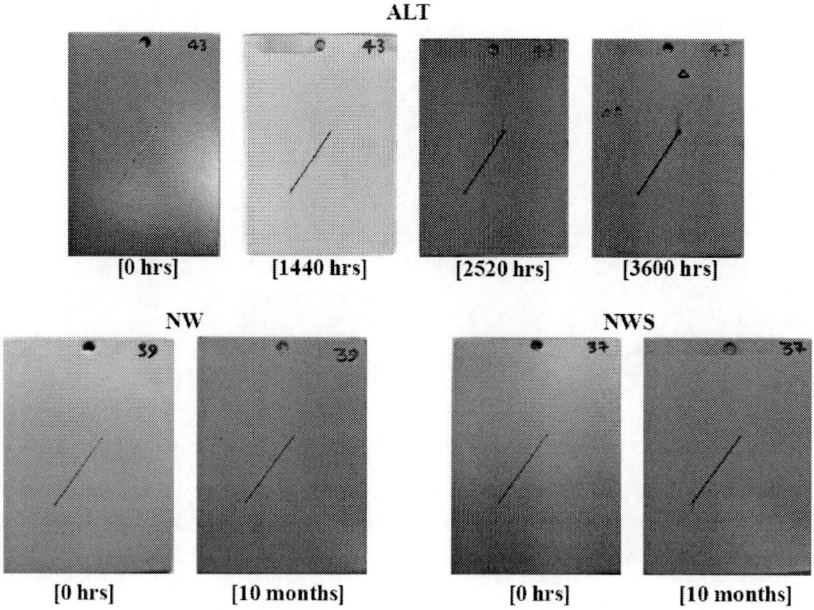

Figure 26. Photo. Progressive changes of ZE/PU—type I in ALT, NW, and NWS.

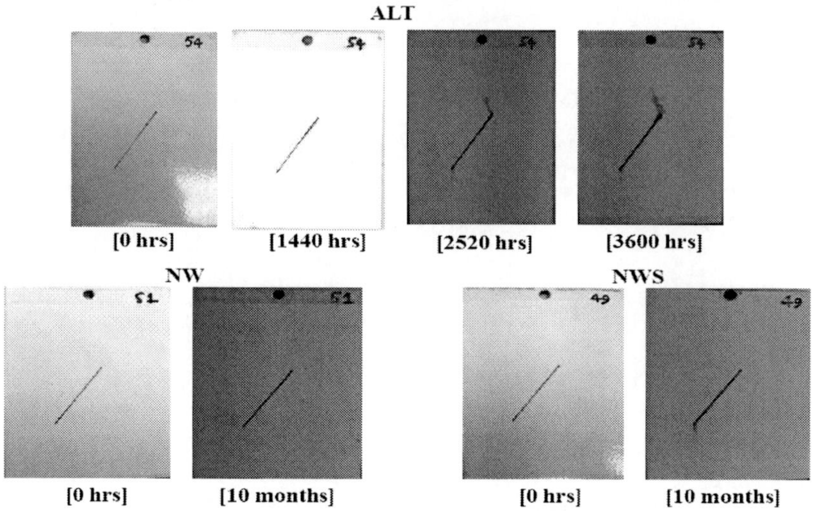

Figure 27. Photo. Progressive changes of Zn/PS—type I in ALT, NW, and NWS.

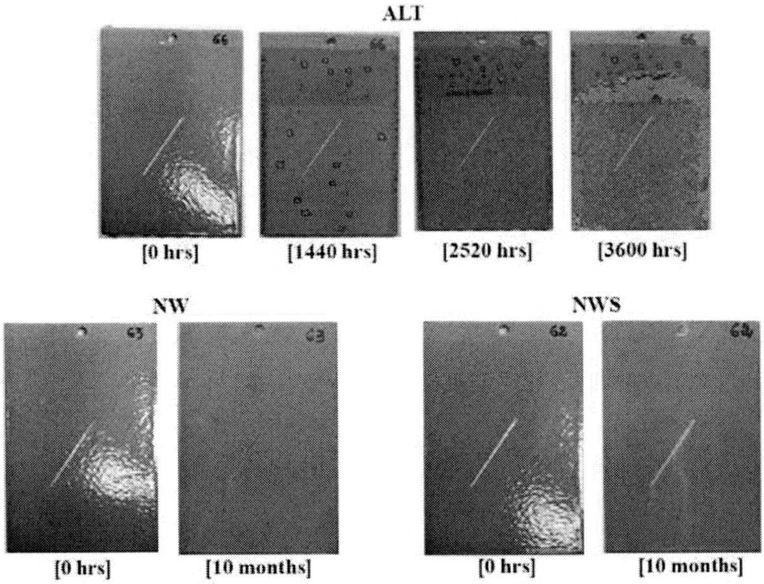

Figure 28. Photo. Progressive changes of TSZ/LE—type I in ALT, NW, and NWS.

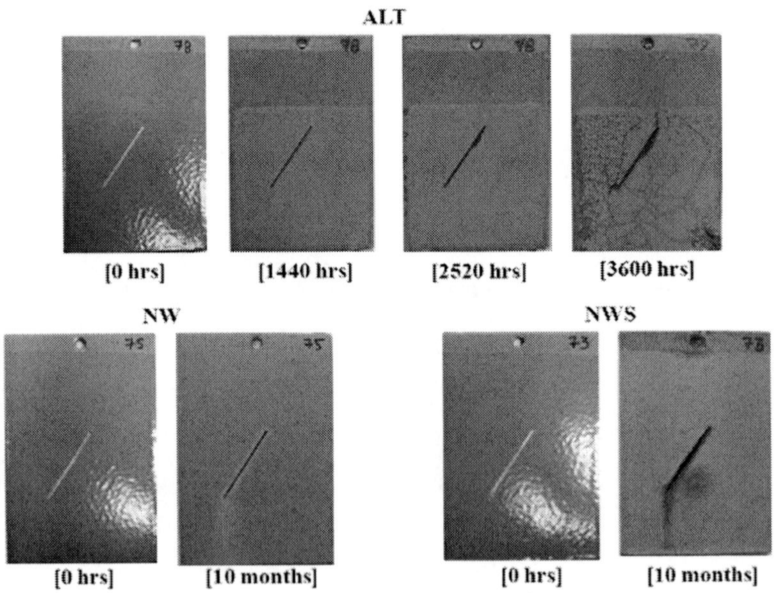

Figure 29. Photo. Progressive changes of ZnE/LE—type I in ALT, NW, and NWS.

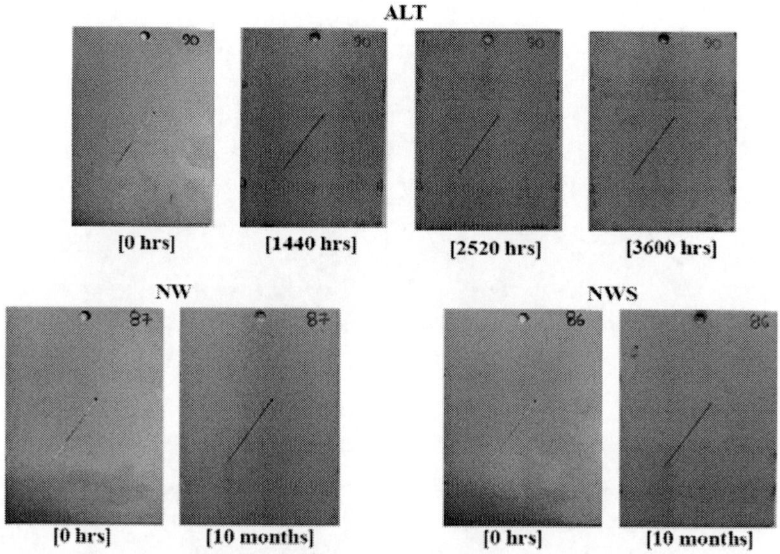

Figure 30. Photo. Progressive changes of HRCSA—type I in ALT, NW, and NWS.

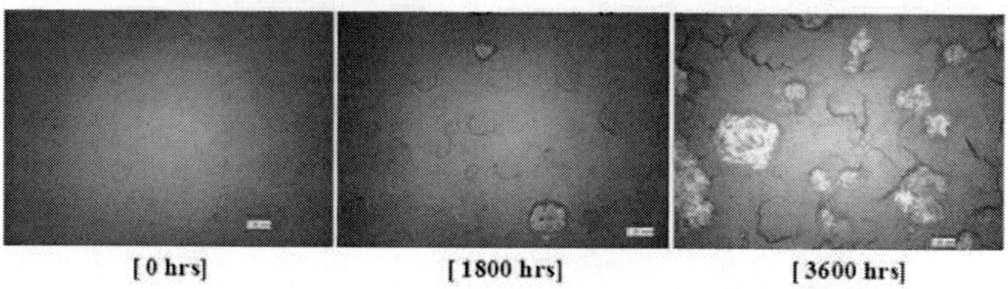

Figure 31. Photo. Photomicrographs of progressive changes of TSZ/LE—type I in ALT.

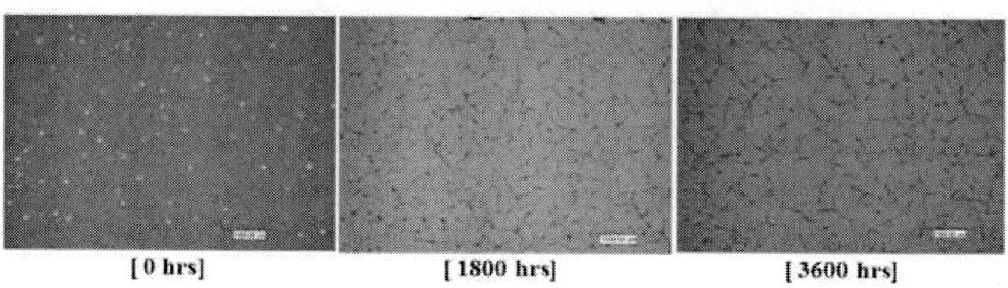

Figure 32. Photo. Photomicrographs of progressive changes of ZnE/LE—type I in ALT.

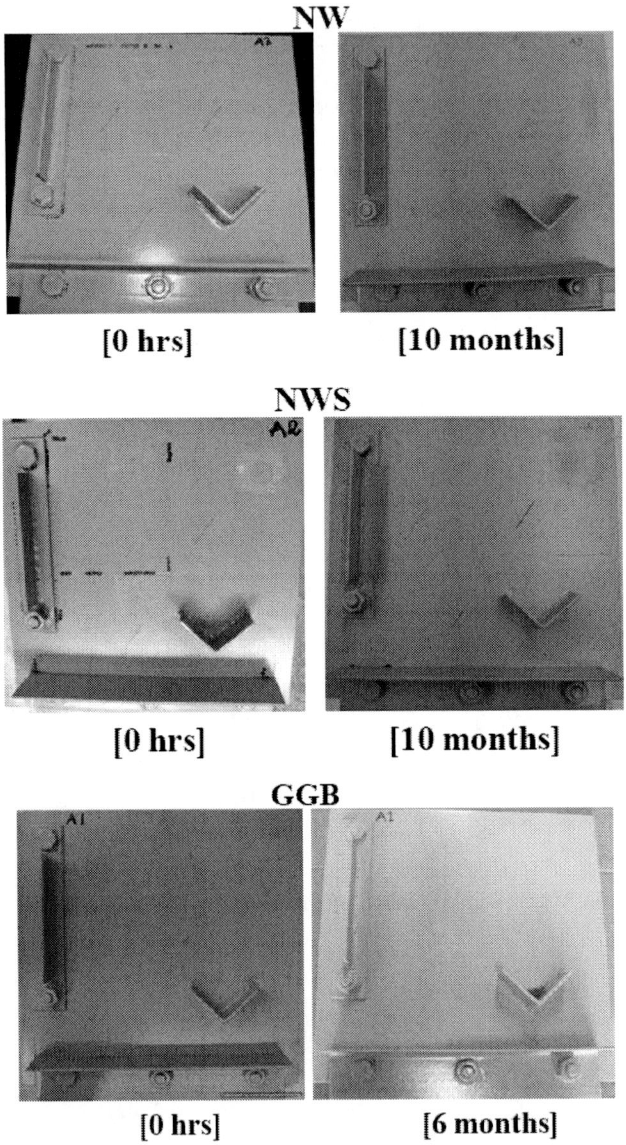

Figure 33. Photo. Progressive changes of IOZ/E/PU—type II in NW, NWS, and GGB.

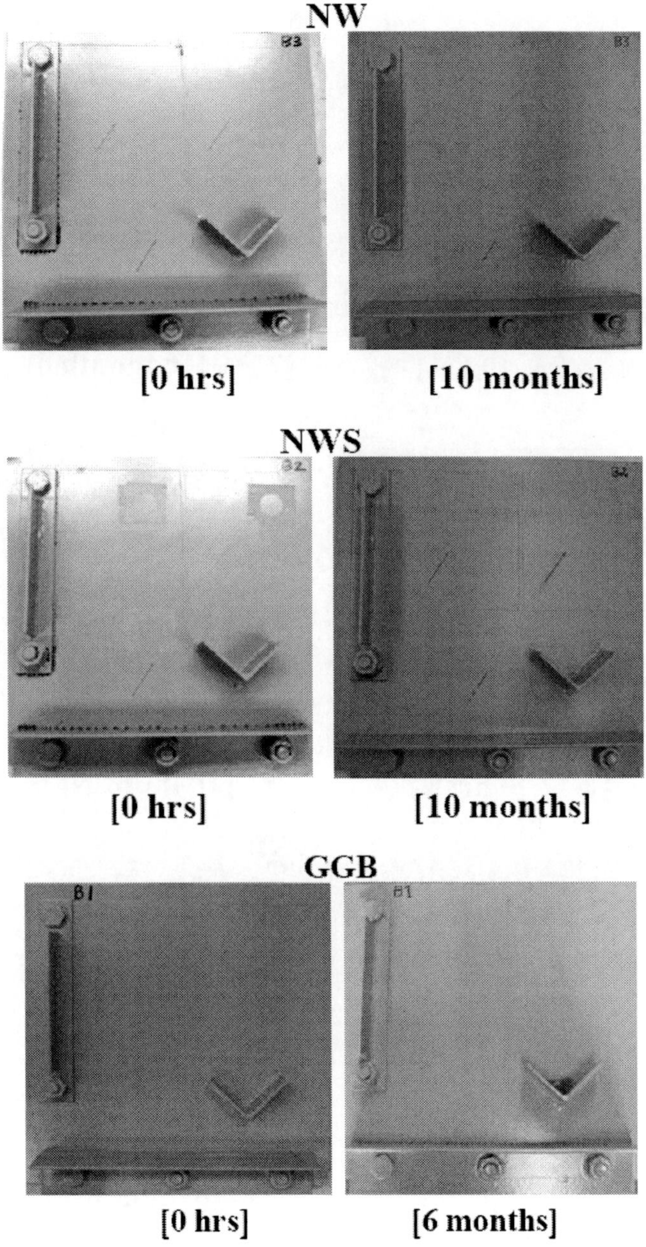

Figure 34. Photo. Progressive changes of ZE/E/PU—type II in NW, NWS, and GGB.

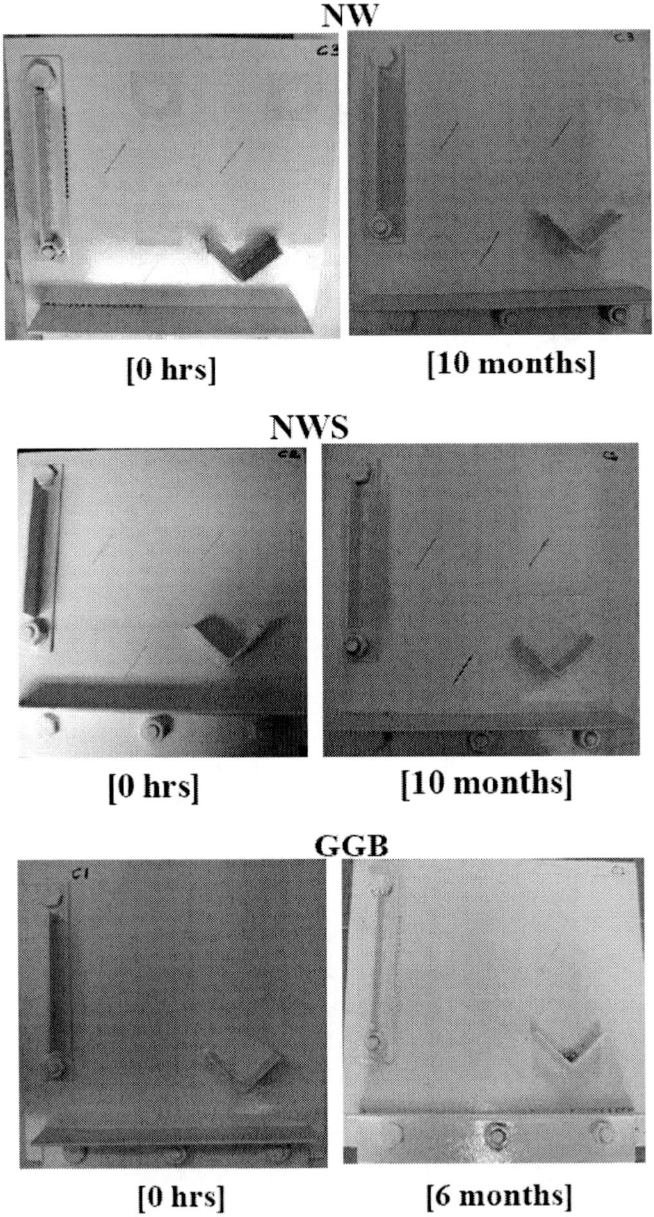

Figure 35. Photo. Progressive changes of MCU/E/F—type II in NW, NWS, and GGB.

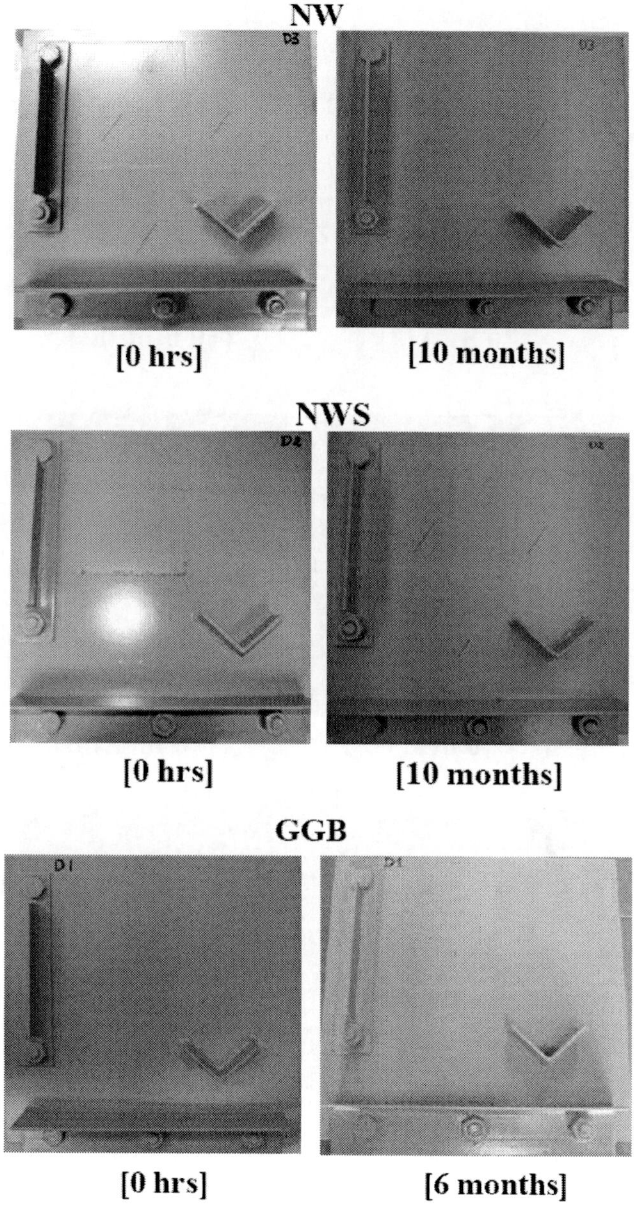

Figure 36. Photo. Progressive changes of ZE/PU—type II in NW, NWS, and GGB.

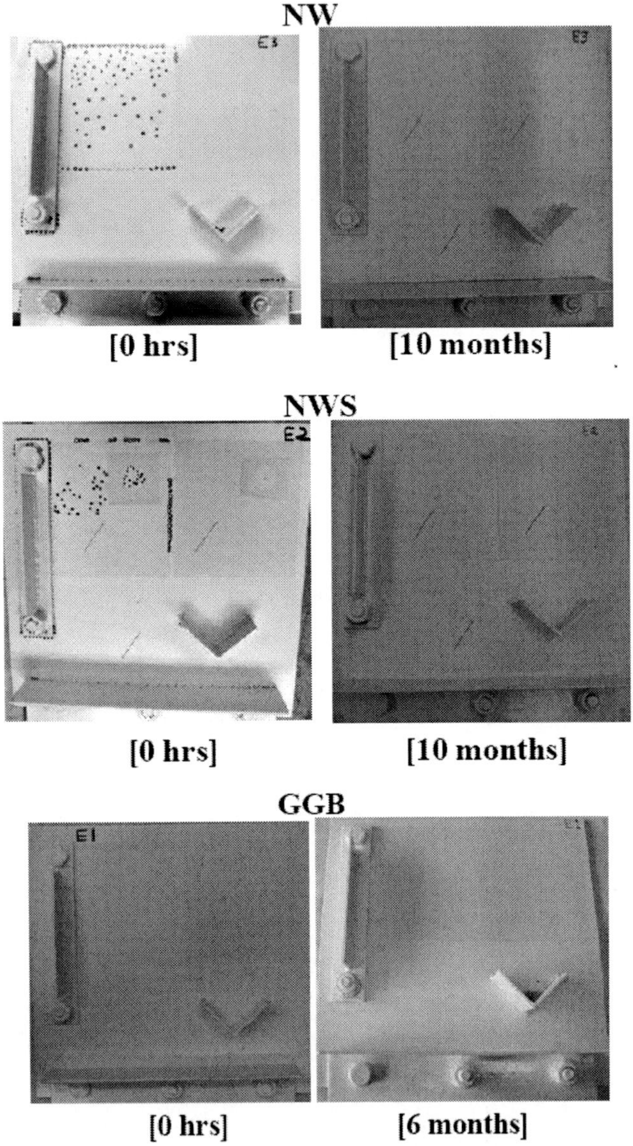

Figure 37. Photo. Progressive changes of Zn/PS—type II in NW, NWS, and GGB.

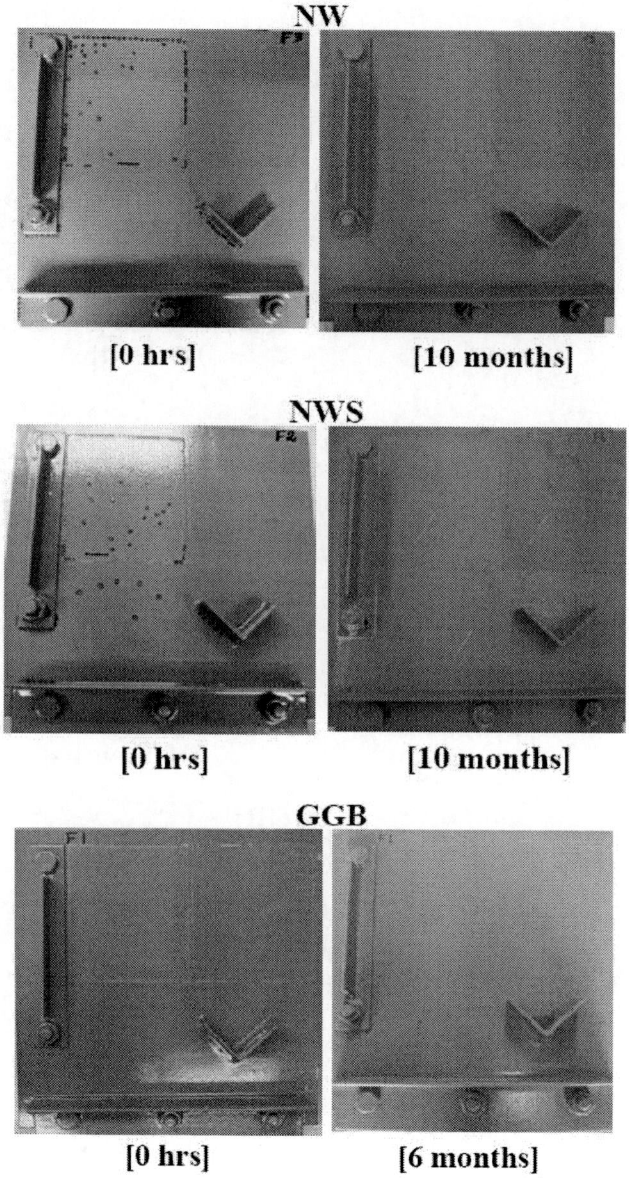

Figure 38. Photo. Progressive changes of TSZ/LE—type II in NW, NWS, and GGB.

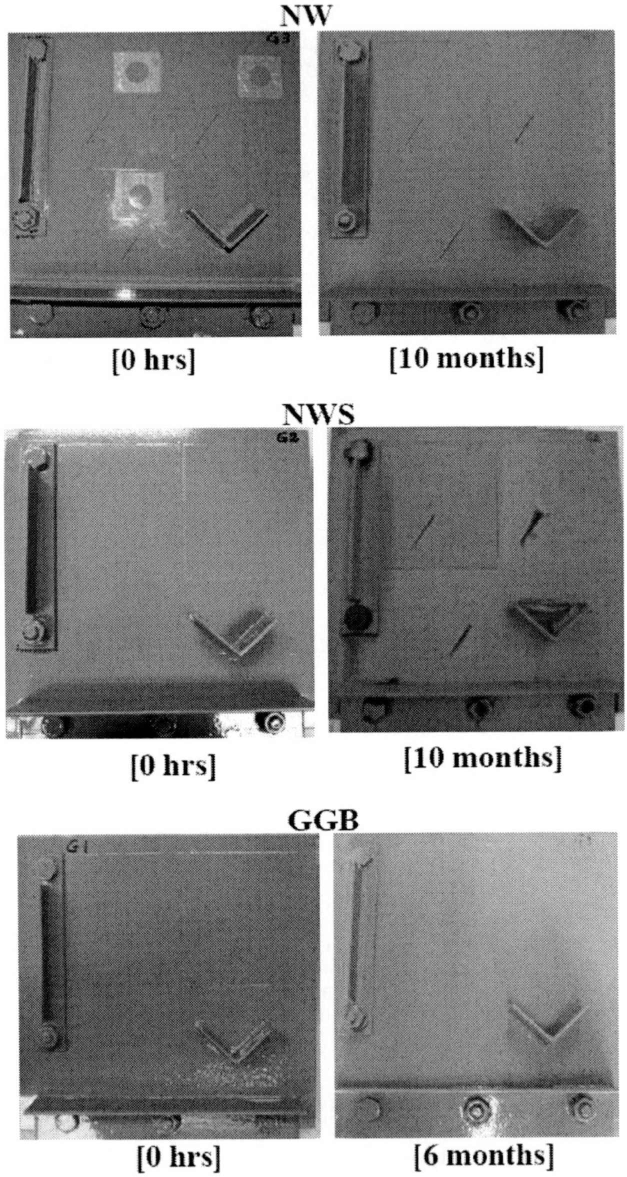

Figure 39. Photo. Progressive changes of ZnE/LE—type II in NW, NWS, and GGB.

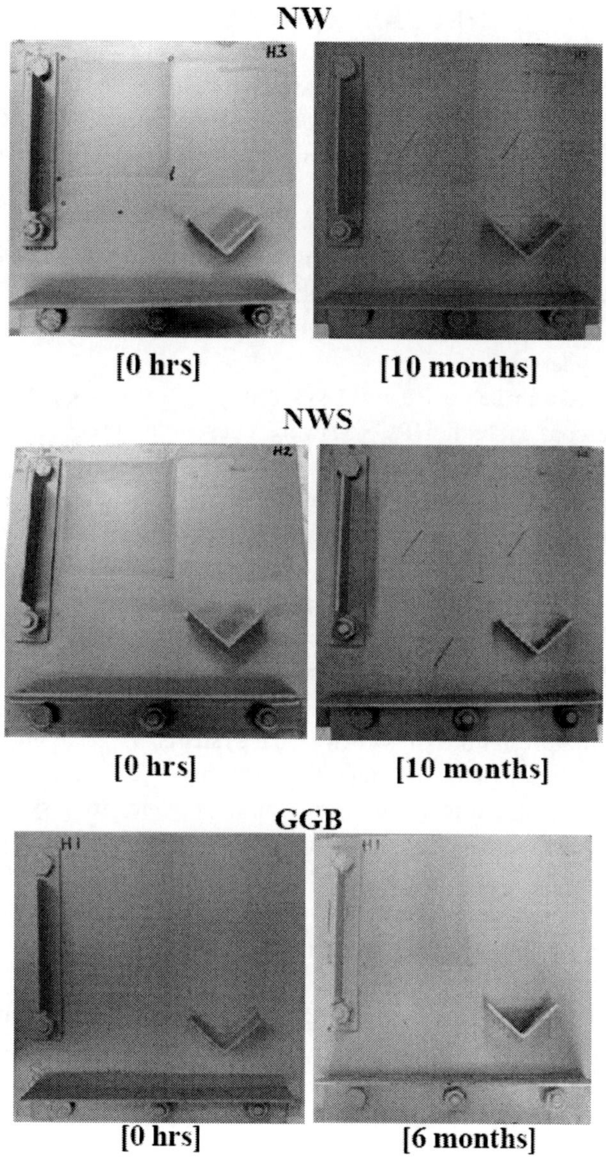

Figure 40. Photo. Progressive changes of HRCSA—type II in NW, NWS, and GGB.

Section 4. Conclusion

The following conclusions are based on test results of the eight coating systems collected during the initial coating characterization, 3,600 h of ALT, 10 months of NW and NWS exposure testing, and 6 months of outdoor exposure testing at GGB:

- Test results from this study indicate that none of the selected coating systems, including the two three-coat control coatings, will provide maintenance-free corrosion protection to steel bridge structures for 100 years.
- The two three-coat control systems, IOZ/E/PU and ZE/E/PU, and the one-coat system, HRCSA, were chosen for their good performance records in an earlier FHWA one-coat study. [22] As expected, they performed well, and they were better than the other test coating systems in every category. The remaining five coating systems, ZE/PU, Zn/PS, TSZ/LE, ZnE/LE, and HRCSA, were selected for a possibility of providing superior performance to commercially available products in the current market. However, they did not deliver desirable performance exceeding the three best coating systems.
- Premature failure of two two-coat systems, TSZ/LE and ZnE/LE, was not anticipated during the coating selection process. Their performance was the worst among the eight coating systems and had a negative impact on this study, leading to early termination of the research. Three test coating systems, MCU/E/F, ZE/PU, and Zn/PS, performed satisfactorily in some categories and poorly in the others compared to the best performers. None of them showed consistently good performance.
- It is apparent that cutting-edge coating technology, regardless of cost, is not ready to deliver super durable coating systems that can last more than 100 years without significant maintenance interventions.
- Until future research and development efforts produce coating systems with extended service life, the main goal should be to use the proven legacy coating systems correctly by reducing human errors and improper applications. At the same time, researchers should strive to develop surface-tolerant primers against salts residue, adhered rusts, and mill scale; a simple yet reliable in situ test method for surface chloride concentration; and allowable chloride

contamination(s) on the blasted steel surface. Significant advancement in these areas will allow for the creation of more durable steel bridge coatings than what is currently available.

REFERENCES

[1] Kogler, B. (2008). "Managing the Infrastructure: The Role of Cost Knowledge," *Journal of Protective Coatings and Linings*, 22–31.

[2] Kline, E.S. (2008). "Steel Bridges: Corrosion Protection for 100 Years," *Journal of Protective Coatings and Linings*, 20–31.

[3] Appleman, B.R. and Kline, E.S. (2003). *Analysis of Cost for Shop Painting of Bridge*, KTATator Report to FHWA.

[4] Chong, S.L. and Yao, Y. (2000). *Laboratory and Test-Site Testing of Moisture-Cured Urethanes on Steel in Salt-rich Environment*, Report No. FHWA-RD-00-156, Federal Highway Administration, Washington, DC.

[5] Chong, S.L. and Yao, Y. (2006). "Performance of Testing of Two-Coat Zinc-Rich Systems on Steel Bridges," *Journal of Protective Coatings and Linings*, 72–83.

[6] Lee, S.K., Chong, S.L., and Yao, Y. (2006). *Performance Evaluation of Bridge Overcoating Materials Using Electrochemical Impedance Spectroscopy*, Proceedings of Paint and Coatings Expo, Tampa, FL.

[7] Chong, S.L. and Yao, Y. (2006). "Selecting Overcoats For Bridges," *Public Roads*, *71*(2).

[8] Lee, S.K., Kogler, B., and Yao, Y. (2010). *Outdoor Performance of One-Coat Systems Applicable to New Steel Bridges*, Proceeding of Paint and Coatings Expo, Phoenix, AZ.

[9] Chong, S.L. and Yao, Y. (2006). "Are Two Coat as Effective as Three?," *Public Roads*, *70*(2).

[10] Society for Protective Coatings. (2007). *Joint Surface Preparation Standard SSPCSP5/NACE No.1: White Metal Blast Cleaning*, Pittsburgh, PA.

[11] ASTM D1654-08. (2008). "Standard Test Method for Evaluation of Painted or Coated Specimens Subjected to Corrosive Environments," *ASTM Book of Standards Volume 06.01*, ASTM International, West Conshohocken, PA.

[12] ASTM G85-11. (2011). "Standard Practice for Modified Salt Spray (Fog) Testing," *ASTM Book of Standards Volume 06.01*, ASTM International, West Conshohocken, PA.

[13] Society for Protective Coatings. (2009). *SSPC-PA2: Measurement of Dry Coat Thickness with Magnetic Gages*, Pittsburgh, PA.

[14] Nadal, M.E. (2001). "NIST Reference Goniophotometer for Specular Gloss Measurements," *Journal of Coatings Technology*, *73*(917), 73–80.

[15] ASTM D523-08. (2008). "Standard Test Method for Specular Gloss," *ASTM Book of Standards Volume 06.01*, ASTM International, West Conshohocken, PA.

[16] ASTM D2244-05. (2005). "Standard Practice for Calculation of Color Tolerances and Color Differences from Instrumentally Measured Color Coordinates," *ASTM Book of Standards Volume 06.01*, ASTM International, West Conshohocken, PA.

[17] ASTM D4541-09e1. (2009). "Standard Test Method for Pull-Off Strength of Coatings Using Portable Adhesion Testers," *ASTM Book of Standards Volume 06.02*, ASTM International, West Conshohocken, PA.

[18] ASTM G62-07. (2009). "Standard Test Methods for Holiday Detection in Pipeline Coatings (Method A)," *ASTM Book of Standards Volume 06.02*, ASTM International, West Conshohocken, PA.

[19] ASTM D714-02. (2009). "Standard Test Method for Evaluating Degree of Blistering of Paints," *ASTM Book of Standards Volume 06.01*, ASTM International, West Conshohocken, PA.

[20] ASTM D610-01. (2009)."Standard Test Method for Evaluating Degree of Rusting on Painted Steel Surfaces," *ASTM Book of Standards Volume 06.01*, ASTM International, West Conshohocken, PA.

[21] ASTM D7087-05a. (2005). "Standard Test Method for An Imaging Technique to Measure Rust Creepage at Scribe on Coated Test Panels Subjected to Corrosive Environments," *ASTM Book of Standards Volume 06.01*, ASTM International, West Conshohocken, PA.

[22] Yao, Y., Kodumuri, P., and Lee, S.K. (2011). *Performance Evaluation of One-Coat Systems for New Steel Bridges*, FHWA Report No. FHWA-HRT-11-046, Federal Highway Administration, Washington, DC.

In: Durable Coating Systems for Steel Bridges ISBN: 978-1-63117-158-1
Editor: Margaret S. Criswell

Chapter 2

PERFORMANCE EVALUATION OF ONE-COAT SYSTEMS FOR NEW STEEL BRIDGES*

Yuan Yao, Pradeep Kodumuri and Seung-Kyoung Lee

ABSTRACT

In an effort to address cost issues associated with shop application of conventional three-coat systems, the Federal Highway Administration completed a study to investigate the performance of eight one-coat systems and two control coatings for corrosion protection of highway bridges. Based on prior performance, a three-coat system and a two-coat system were selected as the control coating systems. The performance of all coating systems was evaluated under accelerated laboratory and outdoor exposure conditions. Accelerated testing was performed in the laboratory for 6,840 h. Natural weathering exposure was performed in the outdoor environment for 18 months and at a marine exposure site for 24 months. A calcium sulfonate alkyd coating system was found to perform equally in comparison with the three-coat system; however, curing was a major concern. Regression analysis was used to identify correlations between color, gloss, adhesion strength, and coating defects for one-coat systems.

* This is an edited, reformatted and augmented version of the Federal Highway Administration, FHWA-HRT-11-046, dated June 2011.

FOREWORD

The current state of art for corrosion protection of steel bridges involves a three-coat system typically consisting of a zinc-rich primer, an intermediate coat, and a top coat. Replacing a three-coat system with fewer coats without sacrificing corrosion resistance can lead to savings in production cost and improved productivity in the steel shops. Two-coat systems evaluated in a previous Federal Highway Administration study performed on par with the widely established well-performing zinc-rich three-coat systems. In this in-house one-coat study, commercially available coating materials that can be applied as one-coat systems to new steel bridges were evaluated. Eight one-coat systems and two controls, a three-coat and a two-coat system, were chosen, and their performance was evaluated using accelerated laboratory testing (ALT) and two outdoor exposure conditions, natural weathering (NW) and natural weathering with salt spray (NWS). ALT (6,840 h) and the two outdoor exposure conditions (NW and NWS), both 18 months, were performed at the Coatings and Corrosion Laboratory at the Turner-Fairbank Highway Research Center in McLean, VA, while another outdoor exposure was performed at a marine exposure site for 24 months in Sea Isle City, NJ. The purpose of this study was to evaluate the performance ranking of various one-coat test systems.

Jorge E. Pagán-Ortiz
Director, Office of Infrastructure Research and Development

SECTION 1. INTRODUCTION

Identification of health hazards associated with lead-based paints in the 1970s led to their replacement with three-coat systems to protect steel bridges from corrosion.[1] Bridge coating technology has been vastly redefined in the past 30 years by changes in surface preparation methodologies, coating processes, and coating material science. Technological advancement in these areas has aided in creating high-quality bridge coating systems with enhanced corrosion protection and minimal environmental impact.

The current state of practice in bridge coatings usually involves multilayer coating typically consisting of a zinc-rich primer over an abrasive blast-cleaned surface and two additional coating layers on top of the primer. The inorganic or organic zinc-rich primer provides cathodic protection by sacrificing itself to the less electrochemically active steel substrate in the presence of corrosive conditions. The intermediate coat provides a physical barrier to the passage of moisture, oxygen, and electrolytes, while the top coat protects against deterioration caused by ultraviolet (UV) radiation while enhancing the aesthetics of the coating. Conventional three-coat systems have demonstrated a long-term service life. Studies have shown that these three-coat systems with a zinc-rich primer can have a service life of 30 years before a major touch-up is required.[2]

Although current coating technology provides a comprehensive solution to improve corrosion protection of steel bridges, the overall cost involved is relatively higher than its predecessors. Data obtained from 20 fabrication shops in the United States for a recent Federal Highway Administration (FHWA)-sponsored study to investigate and analyze the cost of shop painting indicated that the painting cost of steel bridges ranged from under 4 percent to more than 24 percent of the cost of fabricating the steel.[3] The median cost of application of a one-coat system is 8 percent of the cost of the girder, while the cost of a three-coat system is 12 percent for the same application.[3] These increased costs can be attributed to enhanced preprocessing steps, such as higher levels of cleaning and surface preparation, and the direct influence of the number of protective coats on the overall cost of coating fabrication. In addition to the cost involved, the time and space required for proper shop application of a three-coat system are a burden to fabricators and bridge owners. Optimizing cost and productivity is a major challenge for the bridge-coating industry.

In an effort to minimize fabrication costs, novel fast deployment two-coat systems were studied in an FHWA project in 2002. Test results from surface failure and rust creepage at the scribe in both the laboratory test and outdoor exposure revealed that the two-coat systems performed on par with the widely established well-performing zinc-rich three-coat systems. While they are cost-effective due to fewer coats, these two-coat systems have the potential to replace the conventional three-coat systems without sacrificing much corrosion resistance. [4]

Since the performance evaluation of two-coat systems demonstrated promising potential to replace three-coat systems, FHWA sponsored a small research project to investigate the viability of one-coat systems. Three one-coat materials, including a polyaspartic (ASP), a polysiloxane (SLX), and a waterborne epoxy (WBEP), were evaluated in this study.[5] All three one-coat systems developed severe blistering along and away from the scribe area after 5,000 h of salt fog exposure according to American Society for Testing Materials (ASTM) B117-09, "Practice for Operating Salt Spray (Fog) Apparatus."[6] Two of the three one-coat systems did not blister (away from the scribe) after about 5,000 h of cyclic weathering exposure according to ASTM D5894-05, "Standard Practice for Cyclic Salt/Fog/UV Exposure of Painted Metal (Alternating Exposures in a Fog/Dry Cabinet and a UV/Condensation Cabinet)."[7] This may be due to continuous salt fog exposure in ASMT B117-09 compared to cyclic salt fog in combination with UV exposure conditions in ASTM D5894-05.[6,7] Although none of the one-coat systems performed as well as a standard three-coat system, two one-coat systems showed encouraging performance characteristics such as strong adhesion, edge retention, and minimal to no surface blistering in the cyclic weathering test.

In light of the encouraging results obtained from the 2002 study, FHWA performed extensive one-coat research at FHWA's Turner-Fairbank Highway Research Center (TFHRC) Coatings and Corrosion Laboratory (CCL) in McLean, VA. The purpose of this study was to evaluate the performance characteristics of various commercially available high-performance coating materials that can be applied as one-coat systems to steel bridges in shop application. Eight one-coat systems were selected based on their performance in previous FHWA research projects and also after researching the *North East Protective Coat Qualified Products List* and many commercially available coating products. [8]

A three-coat and a two-coat system, both consisting of zinc-rich based primers, were included in this study as controls. The 10 selected coating systems were tested using the cyclic testing method ASTM D5894-05 in addition to a freeze cycle, an accelerated laboratory test (ALT) for 6,840 h, and three outdoor exposure conditions including a marine exposure (ME) in Sea Isle City, NJ, for 24 months, mild natural weathering (NW) for 18 months at TFHRC, and mild natural weathering plus salt solution spray (NWS) tests for 18 months at TFHRC.[7,9]

This report presents performance evaluation results and major findings for the 10 coating systems based on experimental data and subsequent data analyses.

SECTION 2. EXPERIMENTAL PROCEDURES

2.1. Coating Systems

Table 1 lists the 10 coating systems evaluated in this study, including eight one-coat systems, a three-coat system, and a two-coat system. The selected eight one-coat systems are the most popular generic types currently used for steel bridge protection. Individual coatings and the selection criteria are discussed below.

Conventional three-coat systems with epoxy zinc-rich primer, an epoxy intermediate coat, and a polyurethane top coat have been used in steel bridges. They have performed well in the field as well as in previous FHWA studies.[10,11] As a result, a three-coat system was used in this study as one of the two controls.

A two-coat system consisting of a zinc-rich moisture-curing urethane primer and a fast drying aliphatic (AL) polyurea urethane top coat was also selected as the other control because it performed well in a previous FHWA study.[4]

ASP is a new type of coating produced by the reaction of ASP ester compounds (a type of secondary AL amine with AL polyisocyanate).[12] This coating system has a different application procedure and performance properties compared to conventional polyureas. It provides faster drying time, higher film builds than traditional polyurethanes, and high-quality weatherability. ASP can be applied directly to the surface of metal and can provide excellent corrosion protection.

EM, or aluminum-pigmented high solid epoxy coating, has high build in one coat and may have more surface contaminant tolerance while requiring less surface preparation prior to application.[13]

Two waterborne coatings, WBEP and HBAC, were also selected for their low flammability, low odor, and low volatile organic content. Both are fast drying, high film building coating systems.

HRCSA is an alkaline coating that provides excellent corrosion resistance due to its ability to neutralize acidity by promoting passivity at the steel surface. It is designed to have hydrophobic properties and develop strong ionic bonding with metal substrates. The strong corrosion resistance and high-build properties of HRCSA make it an attractive one-coat candidate. HRCSA exhibited excellent performance as an overcoat system in a previous FHWA study.[13] In this study, HRCSA-coated test panels did not exhibit any rust creepage after 4,000 h of ALT or during 24 months of seaside exposure.

Table 1. Summary of coating systems

System Number	System ID	Coating Type		
		Primer	**Intermediate**	**Top**
1	Three-coat	Zinc-rich epoxy	Epoxy	Polyurethane
2	Two-coat	Zinc-rich moisture-curing urethane		ASP
3	One-coat[1]	ASP		
4		Epoxy mastic (EM)		
5		High-ratio calcium sulfonate alkyd (HRCSA)		
6		Glass flake reinforced polyester (GFP)		
7		High-build waterborne acrylic (HBAC)		
8		WBEP		
9		SLX		
10		Urethane mastic (UM)		

[1] One-coat systems contain one coat of paint that acts as the primer/top coat and do not contain an intermediate coat.

Note: The blank cell indicates that the two-coat system does not contain an intermediate layer.

GFP coatings are fast drying, high-build coatings with promising long-term corrosion protection. They have highly desirable properties because glass flakes reduce shrinkage, increase mechanical strength and water resistance, and possess strong chemical resistance.

A relatively new type of coating is SLX, which is typically designed as an organic-inorganic siloxane hybrid binder. SLX typically consists of inorganic silicon-oxygen (Si-O) groups in combination with organic binders. The Si-O group provides excellent weatherability, while the organic counterpart provides corrosion resistance and durability.[14] In an earlier FHWA coating study, a similar type of organic-inorganic hybrid SLX was tested as the top coat of a waterborne inorganic zinc-rich primer, which demonstrated outstanding performance in both laboratory tests and in ME.[11]

UM is a high-solid, high-build acrylic AL urethane system. This high-performing coating has the advantage of being resistant to water and solvents. The AL feature provides strong UV radiation resistance.

2.2. Preparation of Test Panels

Two sizes of steel test panels were used in this study. The small panels were 4 x 6 x 0.2 inches (10 x 15 x 0.48 cm), while the large panels were 6 x 12 x 0.2 inches (15 x 30 x 0.48 cm) (see figure 1 and figure 2). All test panels were blast cleaned to Society for Protective Coatings (SSPC) specification (SP)-10 with measured anchor profiles in the range of 2.2–2.9 mil (55–72 μm).[15] Coatings were then applied on the cleaned test panels using an airless spray method by a professional coating laboratory.

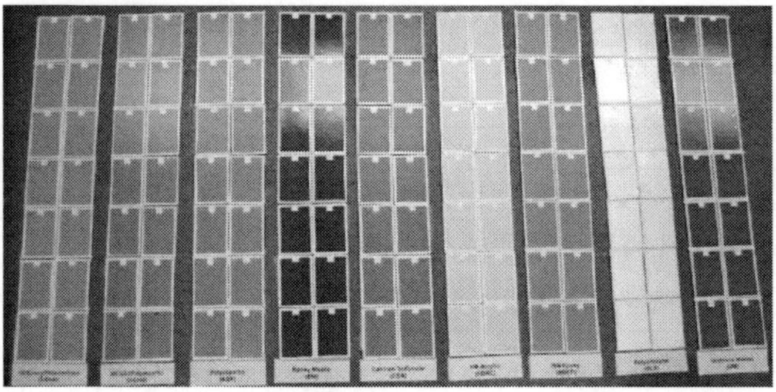

Figure 1. Photo. Composite of small panels.

Figure 2. Photo. Composite of large panels.

Table 2. Test panels used for laboratory and outdoor exposure testing

System Number	System ID	Number of Test Pads						
		ALT (small panels)		ME (small panels)		NW and NWS (large panels)		Total
		Unscribed	Scribed	Unscribed	Scribed	Unscribed	Scribed	
1	Three-coat	3	3	3	3	6	6	24
2	Two-coat	3	3	3	3	6	6	24
3	ASP	3	3	3	3	6	6	24
4	EM	3	3	3	3	6	6	24
5	HRCSA	3	3	3	3	6	6	24
6	GFP	3	3	0	0	0	0	6[1]
7	HBAC	3	3	3	3	6	6	24
8	WBEP	3	3	3	3	6	6	24
9	SLX	3	3	3	3	6	6	24
10	UM	3	3	3	3	6	6	24
Total		30	30	27	27	54	54	222

[1] Indicates that only laboratory test panels were prepared due to the late arrival of coated test panels.

A total of 222 panels were divided into three groups as shown in table 2. The first group consisted of 54 small panels with each coating system covering six panels that were prepared for ALT. Six additional panels coated with GFP were later included in this group for a total of 60 panels. The second group consisted of 54 small panels that were prepared for oceanfront ME in Sea Isle City, NJ. The third group consisted of 108 large steel panels that were prepared for the outdoor exposure racks at TFHRC. Half of the test panels in each group were scribed diagonally following the instructions specified in ASTM D1654-08, "Standard Test Method for Evaluation of Painted or Coated Specimens Subjected to Corrosive Environments."[16] Figure 3 shows a scribing tool being used at TFHRC. The test panels were scribed 2 inches (50.8 mm) long to study the potential performance of the coating systems at local film damage. The other half of the panels were left unscribed to characterize undamaged conditions and physical properties such as gloss, color, pencil scratch hardness, etc. Two additional panels of each coating system were prepared exclusively for initial adhesion strength and Fourier Transform Infrared Spectroscopy Analysis (FTIR) and were not used in any of the tests.

2.3. Test Exposure Conditions

ALT

The cyclic test method described below was employed in ALT. In total, 19 ALT cycles were conducted for a total test period of 6,840 h. This method is similar to ASTM D5894-05, "Standard Practice for Cyclic Salt Fog/UV Exposure of Painted Metal (Alternating Exposures in a Fog/Dry Cabinet and a UV/Condensation Cabinet)," with the addition of a freeze cycle for 24 h.[7]

Each 360-h cycle consisted of the following test conditions in sequence:

1. Freeze for 24 h at -10 °F (-23 °C).
2. Undergo UV/condensation for 168 h (7 days).
   - **Test cycle**: 4 h UV/4 h condensation cycle.
   - **UV lamp**: UVA 1.12 × 10^{-6} ft (340 nm).
   - **UV temperature**: 140 °F (60°C).
   - **Condensation temperature**: 104 °F (40 °C).
3. Undergo prohesion (cyclic salt-fog test, ASTM G85-09, "Standard Practice for Modified Salt Spray (Fog) Testing") for 168 h (7 days).[17]
   - **Test cycle**: 1 h wet/1 h dry.
   - **Wet cycle**: A Harrison mixture of 0.35 weight (wt) percent ammonium sulfate and 0.05 wt percent sodium chloride was used. Fog was introduced at ambient temperature.
   - **Dry cycle**: Air was preheated to 95 °F (35 °C) and then purged to the test chamber.

Figure 4 and figure 5 show a salt fog chamber and a weathering tester, respectively. A 16-h salt-fog accumulation test was conducted before each cyclic salt-fog test to check the atomizing and fog quantity as well as the pH of the collected solution. The test panels were evaluated after every test cycle of 360 h and at the termination of each laboratory test.

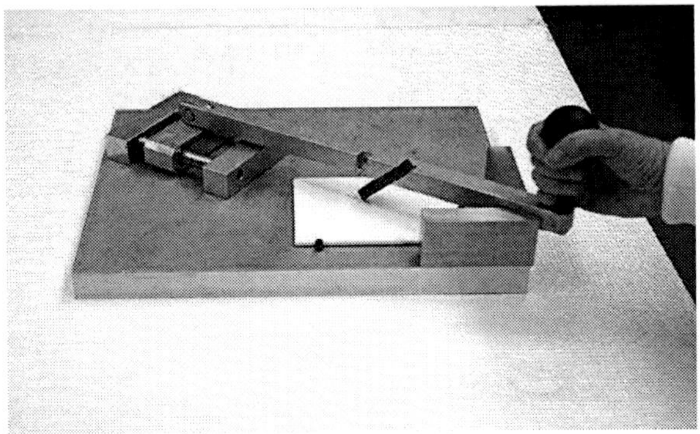

Figure 3. Photo. Example of a scribing tool.

Figure 4. Photo. Salt-fog chamber.

Figure 5. Photo. UV weathering tester.

Oceanfront Exposure

A group of 54 test panels was exposed in Sea Isle City, NJ, at a ME site. All of the test panels were placed on a 45-degree-angle wooden rack facing south as shown in figure 6. The test site is located in a harsh environment with high chloride and high time-of-wetness.[18] Table 3 lists the summary of the environmental factors during the 2-year panel exposure period. The test panels were sent back to TFHRC CCL for performance evaluation every 6 months. The total exposure time was 24 months.

Figure 6. Photo. ME exposure rack in Sea Isle City, NJ.

Table 3. Environmental factors of ME exposure site

Environmental Factors	Measurement
Average seawater temperature (°F)	62
Dissolved oxygen concentration (ppm)	8
Total rainfall (inches)	79
Total hours of sunshine	3,762
Total possible hours of sunshine	6,857
Atmospheric temperature (average high °F)	86
Atmospheric temperature (average low °F)	37
Atmospheric temperature (average °F)	62
Distance of racks from mean high tide (yard)	100
Distance of racks from salt marsh (bayside) (yard)	50

°C = °F-32/1.8.

1 inch = 25.4 mm.

1 yd = 0.914 m.

NW Test Site at TFHRC

A total of 108 panels (12 panels per coating system) were exposed on two racks (54 panels per rack) at the TFHRC weathering test site in McLean, VA, using wooden racks (see figure 7). The test panels were placed at a 30-degree angle on two plastic/wooden racks facing south. The panels on the first rack experienced NW only, and those on the second rack experienced NWS once a day 5 days a week with 15 wt percent sodium chloride solution starting 3 months after initiating the test. The test panels were evaluated every 6 months and at the end of the 18-month test period.

The THFRC exposure site is located in a metropolitan area and was characterized by an annual average precipitation of 32.28 inches (82 cm) in 2007 and 43.31 inches (110 cm) in 2008. Additionally, the annual average salt fall was 0.65 psi (4.48 kPa) chlorine in 2007 and 1.27 psi (8.75 kPa) chlorine in 2008 as measured at the National Oceanic and Atmospheric Administration (NOAA) National Atmospheric Deposition Program/National Trends Network site in Beitsudle, MD. The average annual temperature was 55.4 °F (13°C) in 2007 and 56.5 °F (13.6 °C) in 2008 as measured at a nearby NOAA weather station.

2.4. Coating Characterization Tests and Performance Evaluation Techniques

A series of characterization tests was conducted on the wet coating and dried test panels before ALT and the outdoor exposure tests. The performance of these coating systems in terms of surface defects, rust creepage, and physical and chemical property changes was evaluated during the tests and soon after completion of the tests.

Volatile Content and Solid Content

The volatile and solid content of each of the one-coat systems were obtained following ASTM D2369-10, "Standard Test Method for Volatile Content of Coatings."[19] A sample of certain weight was heated to evaporate volatile components. Volatiles of the coating materials were then calculated from weight loss after heating the raw coating samples. The weight percentage of the solid was calculated by dividing the weight of the solid left after heating by the weight of the coating sample. Two replicates were tested for each coating system, and the mean value was reported.

Figure 7. Photo. Mild NW exposure rack at TFHRC.

Pigment Content

The pigment content of each solvent-based one-coat system was determined following ASTM D2371, "Standard Test Method for Pigment Content of Solvent-Reducible Paints."[20] The pigment fractions were isolated from coating materials by a centrifuge. The solvent used for centrifuge extraction was a mixture of methyl ethyl ketone and toluene in a volume ratio of 1:1. The isolated pigment was then heated to dry. The pigment content was calculated by dividing the pigment weight by the sample weight. The pigment content of two waterborne coatings, HBAC and WBEP, was determined following ASTM D3723-05e1, "Standard Test Method for Pigment Content of Water-Emulsion Paints by Low-Temperature Ashing."[21] The weighed coating sample was added and dispersed in an aluminum dish containing a few milliliters of deionized water. The dish with the sample was initially heated to 221 °F (105 °C) for 1 h followed by another heat treatment at 842 °F (450 °C) for 1 h to burn out the organic binder. The pigment content was then calculated by dividing the solid weight left after heating by the weight of coating sample. Two replicates were tested for each coating system, and the mean value was reported.

Elemental Pigment Analysis

The elemental content of extracted coating pigment in each coating system was analyzed using scanning electron microscopy/energy dispersive x-ray spectrometry technique (SEM/EDS). Characteristic x-rays are produced when a material is bombarded with electrons in SEM. EDS detects the emitted x-rays and converts them into a series of peaks representative of the type and relative amount of each element in the sample. A quantitative analysis method with standards was used to obtain the relative amount of each element in the extracted pigments of the one-coat systems.

These preliminary analyses were conducted to screen the elements present in each of the one- coat systems. A series of standard pellets with material containing the detected elements was made and used as calibration standards. Standard pellets were made by weighing and mixing the chemicals containing the detected elements, using a mortar and pestle, and compacting the mixture powders using a hand press. Test samples were also grinded and pressed into pellets for quantitative analysis of EDS. Figure 8 shows a sample grinding tool and hand press kit.

It is important to note that even after calibration with standards, the amount of each element determined from SEM/EDS analyses was semiquantitative because of the limitation of the EDS technique and the practical limitation of users when making qualitatively similar standards.

FTIR Analysis

FTIR spectra of the test panels were obtained before and after the laboratory testing and outdoor exposure. FTIR produces an absorption spectrum that provides information about the chemical bonds and functional groups present in the molecules. Typically, a FTIR spectrum is equivalent to the "fingerprint" of the material and can be compared with cataloged FTIR spectra to identify the chemical composition of the material.

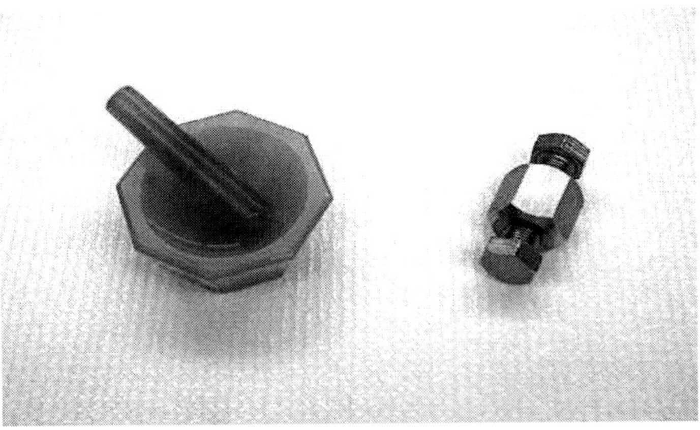

Figure 8. Photo. Simple grinding tool and hand press kit.

Samples for analysis were collected from the surface of coating panels using a knife scratch technique. Single reflection horizontal attenuated total reflectance spectra were obtained using a diamond accessory. Two samples for each coating system were collected from the unscribed test panels. The relative amount of aromatics to aliphatics (AR/AL) of binder was estimated from the ratio of the peak area at a wave number of 3,100 cm^{-1} to that at a wave number of 3,000–2,800 cm^{-1}. The reported AR/AL ratios were the mean value of the data obtained from two replicates. Top coat samples were collected and analyzed for the two-coat and three-coat systems by carefully scratching the top coat layers.

Sag Resistance

Sag resistance of the coatings was determined following ASTM D4400-99, "Standard Test Method for Sag Resistance of Paints Using a Multinotch Applicator (Method A—Horizontal Test Method)."[22] Coating was applied to a test chart with a multinotch applicator and was hung vertically with the drawdown strips horizontal and the thinnest strip at the top. Drawdown was examined after drying, and it was rated for sagging.

Drying Time

Drying time of the coatings was determined using ASTM D1640-03, "Standard Test Methods for Drying, Curing, or Film Formation of Organic Coatings at Room Temperature."[23] Coating with manufacturer-recommended wet film thickness was placed on clean glass plates using a doctor blade. Dry-to-touch time was determined by lightly touching the film using a finger, immediately placing the finger against a piece of clean and clear glass, and examining the glass for any paint that was transferred. The dry-through (or dry-to-handle) time was determined by firmly pressing the film using a thumb and visually inspecting the film to see if it exhibited any wrinkling or physical damage. The sampling plates were placed in a walk-in environmental chamber immediately after the film was placed. The temperature and the relative humidity of the environmental chamber were 77 ±33.8 °F (25 ±1 °C) and 50 ±2 percent, respectively.

Dry Film Thickness (DFT)

DFT of the coatings was measured before ALT and the outdoor exposure tests using an electronic gauge using the SSPC paint application SP-2, "Measurement of Dry Coating Thickness with Magnetic Gauges."[24] Three DFT spot readings were obtained from each one of the test panels. Additionally, three spot readings for each of the small test panels and six spot readings for each of the large test panels were obtained. The reported DFT of each coating system was the mean of the data obtained from all of the test panels.

Gloss

Gloss is the perception of a shiny surface by human eyes. Specular gloss compares the luminous reflectance of a test specimen to that of a standard specimen under the same geometric condition.[25] Measurements by this test method correlate with visual observations of surface shininess made roughly at the corresponding angles. Measured gloss ratings are obtained by comparing the specular reflectance from the specimen to that from a black glass standard. The measured gloss ratings change as the surface refractive index changes because specular reflectance depends on the surface refractive index of the specimen.

Gloss of all of the one-coat system coatings was measured following ASTM D5230-08, "Standard Test Method for Specular Gloss."[26] The 60-degree geometry measurements were conducted on the selected unscribed test panels prior to laboratory and outdoor exposure tests. Three gloss readings for each of the small test panels and six readings for each of the large test panels were recorded. The reported gloss of each coating system per test condition was the mean of the readings obtained from all unscribed test panels.

Color

The color of the coatings was measured using a 45-degree/zero-degree colorimeter following ASTM D2244-09A, "Standard Test Method for Calculation of Color Differences from Instrumentally Measured Color Coordinates."[27] This technique is based on the calculation from instrumentally measured color coordinates based on daylight illumination of color tolerances and small color differences (ΔE) between opaque coated panels. The International Commission on Illumination (CIE) lab color system (CIE (L*, a*, b*)) was used for color measurement. L*, a*, and b* represent the three coordinates of the three-dimensional lab color space. These parameters are defined based on the following high and low values they represent to identify colors:

- L* = 0 represents black, and L* = 100 represents diffuse white.
- Positive values of a* indicate green, and negative values indicate magenta.
- Positive values of b* indicate blue, and negative values indicate yellow.

- The asterisk (*) is used to differentiate the CIE (L*, a*, b*) system from (L, a, b) parameters of the original Hunter 1948 color space.

Colors were measured for unscribed test panels only before and after ALT and the outdoor exposure tests. Three color readings were obtained for each of the small test panels, and six color readings were obtained for each of the large test panels. ΔE of the test panels before and after the test was calculated using the following equation:

$$\Delta E = [(\Delta L^*)^2 + (\Delta a^*)^2 + (\Delta b^*)^2]^{1/2} \quad (1)$$

where:

ΔL^* = L*after test - L*before test.
Δa^* = a* after test - a* before test.
Δb^* = b* after test - b* before test.

The data used in the above equation were the mean of the data obtained from all the unscribed test panels of each coating system.

Pencil Scratch Hardness

The pencil scratch hardness of all coating systems was determined following ASTM D3363-05, "Standard Test Method for Film Hardness by Pencil Test."[28] In this test, a pencil hardness gauge installed with pencils of various grades of hardness was used to scratch the coating film. The grade of the hardest pencil that did not scratch the film was referred to as the pencil scratch hardness of the particular coating system. Two unscribed test panels of each coating system were tested before and after ALT and the outdoor exposure tests. The mean of the two readings was used as the nominal hardness of a coating system.

Adhesion

The adhesion strength of the coating systems was determined using the pull-off adhesion testers following ASTM D4541-09, "Standard Test Method for Pull-Off Strength of Coatings Using Portable Adhesion Testers."[29]A loading fixture, commonly known as a dolly or stub, was affixed to the panel surface by an adhesive. A load provided by the adhesion tester was increasingly applied to the dolly until it was pulled off. The force required to pull the dolly off yielded the tensile strength in pounds per square inch or megapascals. Failure occurs along the weakest plane(s) within the testing system comprised of the dolly, adhesive, individual layers of the coating system, and substrate.

The surface of coated test panels and the base of the dollies were cleaned with detergent water and were lightly roughened with an abrasive pad. The dollies were glued to the test panel surface using a high-strength epoxy adhesive. The cut through the coating around the edge of the dolly was performed using a drill press after the complete cure of the adhesive shown in figure 9. The initial adhesion strength was measured by two pull-off methods: the hydraulic method and the pneumatic method. The final adhesion strengths of coating systems were measured using the hydraulic method. Figure 10 shows the hydraulic adhesion tester used to pull off the dollies for adhesion strength testing. Three pull-off adhesion tests were performed after exposure on each test panel except for some UM and ASP test panels due to their severe surface failures after ALT. No adhesion tests were performed on scribed UM and ASP test panels, while two adhesion tests were performed on unscribed UM panels.

Figure 9. Photo. Drill press to score a test area around a dolly.

Figure 10. Photo. Hydraulic adhesion tester.

For every panel, the average adhesion strength of three locations was calculated. If the coefficient of variance (CV) of each test panel was more than 20 percent, the test panel and adhesion failure mode were carefully examined to see if the variation was caused by test operation. Repeat tests were performed for quality assurance. If more than 50 percent of a glue failure occurred, the test was repeated. The reported adhesion strength for each coating system was the mean of the data obtained from tests conducted on all test panels of the coating system. The remaining DFT at the pull-off spots was measured and recorded. The adhesion failure mode of every spot was also documented using digital photographs.

Detection of Coating Defects

The coating defects were identified using ASTM D5162-08, "Standard Practice for Discontinuity (Holiday) Testing of Nonconductive Protective Coating on Metallic Substrates."[30] This technique utilizes a low voltage holiday detector to determine the presence of electrically conductive coating defects including holidays and pinholes (invisible defects), voids, and metal particles protruding through the coating. The reported number of defects after each test cycle was the cumulative number of defects. In addition to using a holiday detector to determine the number of such defects, test panels were visually examined using ASTM D714-02, "Standard Test Method for Evaluating Degree of Blistering of Paints" and ASTM D610-08, "Standard Test Method for Evaluating Degree of Rusting on Painted Steel Surfaces."[31,32] The reference standards were employed to grade the rust pits and surface blisters on the panels.

Digital Microscopic Examination

When unusual surface failures were detected by the holiday detector, the panels were examined using a stereomicroscope or a high-power digital microscope. The surface conditions were documented via microphotographs.

Digital Photography

Every test panel was photographed to document initial surface conditions before initiating the tests. The test panels were also photographed after each test cycle for both ALT and the outdoor exposure tests.

Rust Creepage Measurement

The rust creepage at the scribe was measured following ASTM D7087-05A, "Standard Test Method for An Imaging Technique to Measure Rust

Creepage at Scribe on Coated Test Panels Subjected to Corrosive Environments."[33] The rust creepage area from the scribe line on the coating panel was traced using a thin marker and a transparent plastic sheet. The tracing image was scanned and analyzed using imaging software to obtain the creepage areas and the creepage distances. Two traces for each test panel were obtained, and the mean creepage distance was reported as the nominal creepage for the coating system.

Electrochemical Impedance Spectroscopy (EIS)

The impedance of the coating systems was measured using an electrochemical instrument equipped with a potentiostat. This technique involves applying a small amplitude alternating current signal into a body of material over a wide range of frequencies and measuring the responding current and its phase angle shift. The output from the EIS instrument is an impedance spectrum of the material, typically ranging from 100 to 0.001 Hz. EIS data are analyzed by the equivalent circuit modeling technique, which can produce appropriate models to evaluate the coating deterioration process. The mechanism of corrosion occurred at the interface between the substrate and the coating.

Linear Regression Analysis

Correlation among test parameters, such as color or gloss, for the tested coating systems can aid in developing one or more relationships that can provide better understanding of interactions among test variables. This correlation would be specific to the type of exposure condition involved such as ALT or outdoor exposure testing.

Another type of relationship that could be understood with numerical correlation is how variation of performance evaluation parameters, such as color or gloss, with time in one exposure condition compares to the variation of the same parameter in another exposure condition. One such example would be to correlate color variation in ALT with color variation in an outdoor exposure condition. This type of correlation can help explain how one exposure condition would compare to or simulate the other.

Regression analysis has been performed on the performance data of all of the one-coat systems to identify and understand the above relationships. Panels with the GFP coating system were not available for outdoor testing; therefore, the GFP system was excluded from regression analysis.

Exposure Conditions

The following exposure conditions were employed to evaluate the performance of the 10 coating systems:

- ALT.
- ME.
- NW.
- NWS.

SECTION 3. RESULTS AND DISSCUSION

3.1. Characterization of Coating Systems

All coating systems before ALT and outdoor exposure tests were characterized for the following properties:

- Volatile and pigment contents.
- Major elemental content (wt percent of extracted pigment).
- FTIR and AR/AL.
- Sag resistance.
- Drying time.
- DFT.

Volatile and Pigment Contents

Table 4 lists the volatile, solid, pigment, and binder contents by wt percentage for all one-coat systems. Five one-coat systems, SLX, EM, HRCSA, ASP, and UM, contained solids greater than 70 wt percent, while the remaining three, HBAC, GFP, and WBEP, had solid content in the range of 56–67 wt percent. The pigment content ranged between 27 and 39 wt percent except for GFP, which had a pigment content of 19 wt percent.

Pigment volume concentration is an important parameter in coating formulation because it affects the coating film properties such as gloss, permeability, and blistering resistance.[34] Table 4 lists the pigment and binder contents in weight percentage. These data can be used as references when selecting coating systems.

Table 4. Volatile, solid, pigment, and binder contents

Parameter (wt percentage)	ASP	EM	HRCSA	GFP	HBAC	WBEP	SLX	UM
Volatile	23	11	23	35	33	43	8	24
Solid	77	89	77	65	67	57	92	76
Pigment	38	39	27	19	27	31	30	29
Binder	39	50	50	46	40	26	62	47

Major Elemental Compositions of Pigment Fractions

Table 5 lists the major elemental contents obtained from SEM/EDS analysis. Zinc, iron, aluminum, phosphorus, titanium, silicon, and calcium were present in almost every one-coat system. Pigments in coatings can usually be divided into three categories.[34] The prime pigments represented by titanium oxide and iron oxide provide opacity, color, and protection of the resin against UV light. Additionally, functional pigments, such as anticorrosive inhibitors, provide corrosion resistance. Zinc, aluminum, ferrous, and calcium in the forms of phosphates, borates, and molybdates are common nontoxic anticorrosive pigments. Metallic pigments, such as aluminum and zinc, are also used as inhibitive pigments. Extender pigments, such as calcium carbonate and silica, are used to build the pigment volume and control the physical properties of the coating film.

All one-coat systems, except for EM, have demonstrated presence of titanium as the prime pigment. ASP, HRCSA, HBAC, WBEP, SLX, and UM contained a significant amount of zinc (13–38 wt percent), phosphorus (2–5 wt percent), and aluminum (3–7 wt percent). Based on the elemental content, it is reasonable to assume that some forms of zinc phosphate and/or aluminum zinc phosphate were the major inhibitive pigments present in these one-coat systems. There was no zinc or phosphorus detected in EM and GFP. Aluminum (57 wt percent) was the major anticorrosive element in EM. GFP had a certain amount of aluminum and iron, both of which are anticorrosive elements. Silicon, as an extender pigment, was present in all of the one-coat systems in various weight percentages (4–37 wt percent). GFP contained the largest amount of silicon (37 wt percent), which was assumed to be from the glass flake used to reinforce the coating. Several one-coat systems contained calcium, which acts as an anticorrosive pigment or extender pigment.

Table 5. Major elemental contents in one-coat systems of extracted pigments

Coating System	Element Content (wt percentage)						
	Aluminum	Silicon	Phosphorus	Calcium	Titanium	Iron	Zinc
ASP	6	13	2	0	5	0	38
EM	57	16	0	0	0	1	0
HRCSA	5	4	4	11	11	1	14
GFP	5	37	0	4	21	7	0
HBAC	3	2	3	8	15	1	23
WBEP	4	21	2	5	10	3	13
SLX	5	2	3	0	44	0	17
UM	7	2	5	0	35	2	15

FTIR

Figure 11 shows a typical FTIR spectrum of the three-coat system with characterization peaks before ALT. FTIR spectra of all one-coat systems before and after ALT were recorded for chemical analysis.

Aromaticity, or presence of AR compounds in a coating system, is indicated by the ratio of FTIR peak area of AR (wave number range of 3,100 to 3,000 cm^{-1}) to AL (wave number range of 3,000 to 2,800 cm^{-1}). This ratio is denoted by (AR/alophaticity (AP)) x 100 in table 6. Presence of AR compounds can result in reduced weathering performance in outdoor exposure conditions since UV light causes modified surface appearance of aromatic coatings due to yellowing and/or chalking. The binder of several coating systems (three-coat, EM, GFP, HBAC, WBEP, SLX, and UM) consisted of some degrees of aromaticity, which typically reduces the weatherability of these coating systems. The AR/AP ratio of all one-coat systems in table 6, when correlated with gloss reduction, demonstrated that higher AR/AP resulted in higher gloss reduction (see figure 78).

No AR peaks were detected in the two-coat control system, ASP, or HRCSA. The epoxy coating systems (EM and WBEP) and GFP had the largest amounts of aromaticity indicated by AR/AP greater than 4.5. UM, SLX, the three-coat system, and HBAC were more AL in nature, as indicated by their AR/AP ratios between 1.5 and 2.5.

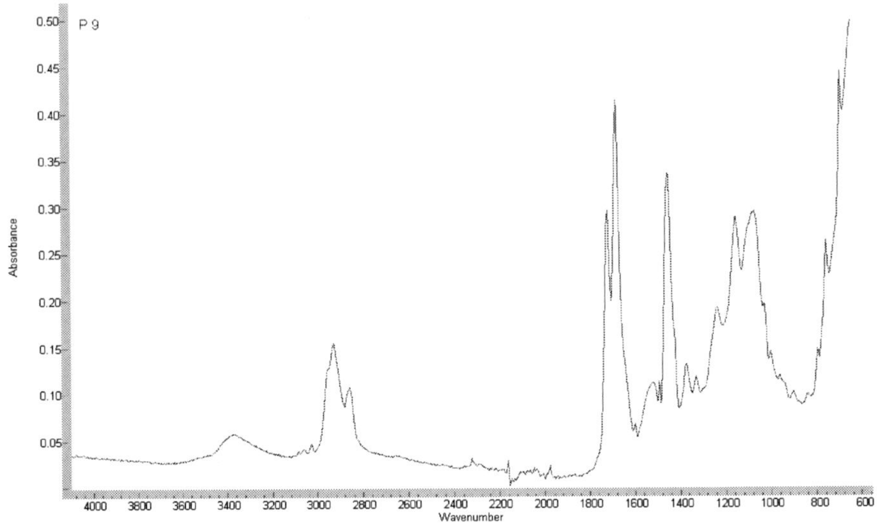

Figure 11. Graph. FTIR spectrum of three-coat system (top coat) before ALT.

Table 6. FTIR AR/AP peak ratio of coating systems

Coating System	AR/AP x 100
Three-coat	2.3
Two-coat	0
ASP	0
EM	5.2
HRCSA	0
GFP	4.5
HBAC	2.5
WBEP	4.5
SLX	1.5
UM	1.5

There were no significant differences between the FTIR spectra obtained before and after ALT. Several spectra of tested panels indicated a few small peaks and some degrees of resolution at certain wave numbers, which may be attributed to possible coating deterioration. The differences in spectra before and after ALT were predominant for the two epoxy coating systems (EM and WBEP) in comparison to the others.

Sag Resistance

SLX sagged at a thickness of 10 mil (254 μm), while UM sagged at 7 mil (177.8 μm). Wet film thickness values for the coating systems were calculated based on DFTs and solid content. The highest wet film thicknesses recommended by the manufacturers were 7.8 mil (198.12 μm) for SLX and 7.9 mil (200.66 μm) for UM, respectively. SLX did not sag at the specified wet film thickness; however, UM sagged at the high end of the manufacturer-recommended wet film thickness. The horizontal antisag indexes of these two systems are shown in table 7. All other coating systems did not sag even at 24 mil (609.6 μm), indicating that these systems had good sag resistance.

Drying Time

Drying time is an important coating property because slow drying coatings lower the productivity in shop applications. In the field, slow drying coatings delay inspections. Table 8 lists the mean dry-to-touch time and the dry-through (dry-to-handle) time of one-coat systems obtained at 77 ±35.6 °F (25 ±2 °C) and 50 ±2 percent relative humidity. Except for HRCSA, all one-coat systems are considered fast drying systems. The two waterborne coating systems, HBAC and WBEP, were the fastest drying one-coat systems with dry-to-touch times of 0.7 h or less and a dry-through time of 3.6 h or less.

HRCSA had the longest dry-to-touch time of 48 h, and it had not reached the status of dry-to- handle even after 240 h of testing. The long set-to-touch time and the long dry-through time can be a serious field drawback, and such slow drying time should be considered prior to application.

Table 7. Horizontal antisag index of SLX and UM

Antisag Parameter	Coating System	
	SLX	UM
Index-stripe number	8	6
Post-index stripe	10	7
Addendum fraction	0	0.6
Index addendum	0	0.6
Antisag index	8	6.6

Table 8. Mean drying time

Coating System	Dry-to-touch Time (hours)	Dry-through Time (hours)
ASP	2.5	5
EM	2.8	5
HRCSA	48	>240
GFP	1.7	3.2
HBAC	0.7	3.5
WBEP	0.5	3.6
SLX	3.8	6.3
UM	3.8	9

DFT

The initial DFT of the one-coat systems and the two controls are listed in table 9. The standard deviation and CV are also listed. The measured DFTs were within the range of the manufacturer- recommended target DFTs except for HBAC, which was about 3 mil (76.2 μm) thicker than the manufacturer-recommend DFT. Figure 12 shows the plot of the DFT data.

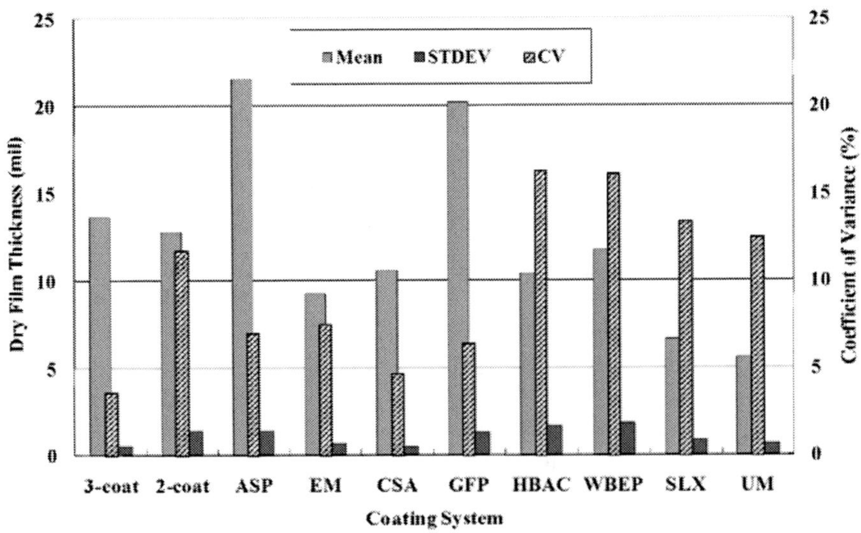

1 mil = 25.4 μm.

Figure 12. Graph. DFT data for the 10 coating systems.

Table 9. Mean DFT

Coating System	Mean (mil)	Standard Deviation (mil)	CV (percent)
Three-coat	13.7	0.5	3.6
Two-coat	12.8	1.5	11.7
ASP	21.5	1.5	7.0
EM	9.3	0.7	7.5
HRCSA	10.6	0.5	4.7
GFP	20.2	1.3	6.4
HBAC	10.4	1.7	16.3
WBEP	11.8	1.9	16.1
SLX	6.7	0.9	13.4
UM	5.6	0.7	12.5

1 mil = 25.4 μm.

The three-coat system had a DFT of 13.7 mil (347.98 μm), and the two-coat control had a DFT of 12.8 mil (325.12 μm). The DFT of the one-coat systems varied significantly: ASP and GFP had the highest DFTs with an average around 20 mil (508 μm). UM had the lowest DFT of 5.6 mil (142.24 μm), followed by SLX with a DFT of 6.7 mil (170.18 μm). The other one-coat systems had DFTs near 10 mil (254 μm). It should be noted that even though UM had the thinnest DFT, as specified in its product data sheet, this coating system developed many surface blisters and rust pits after 4,320 h of ALT in addition to significant rust creepage at the scribe. These poor performance indicators could be attributed to the insufficient DFT value.

3.2. ALT and Outdoor Exposure Testing

Performance of the test coating systems in ALT, ME, NW, and NWS was evaluated using the following parameters:

- Gloss reduction.
- Change of color.
- Change of pencil scratch hardness.
- Change of adhesion strength.
- Development of surface defects and holidays.
- Growth of rust creepage at the scribe.

Gloss Reduction

Overall gloss reduction values are summarized in table 10 and shown in the graph in figure 13.

Table 10. Summary of mean gloss reduction data

Coating System	Exposure Condition (percent)			
	ALT	ME	NW	NWS
Three-coat	50.9	28.9	29.5	14.2
Two-coat	60.3	91.5	39.0	34.5
ASP	27.6	52.7	10.1	15.0
EM	99.0	97.7	96.9	97.3
HRCSA	66.7	30.6	81.9	74.1
GFP	41.6			
HBAC	79.5	29.2	24.4	16.5
WBEP	77.8	66.9	59.3	63.8
SLX	18.5	32.8	20.6	12.4
UM	23.8	4.3	1.5	0.5

Note: The blank cells indicate that no outdoor exposure data were available.

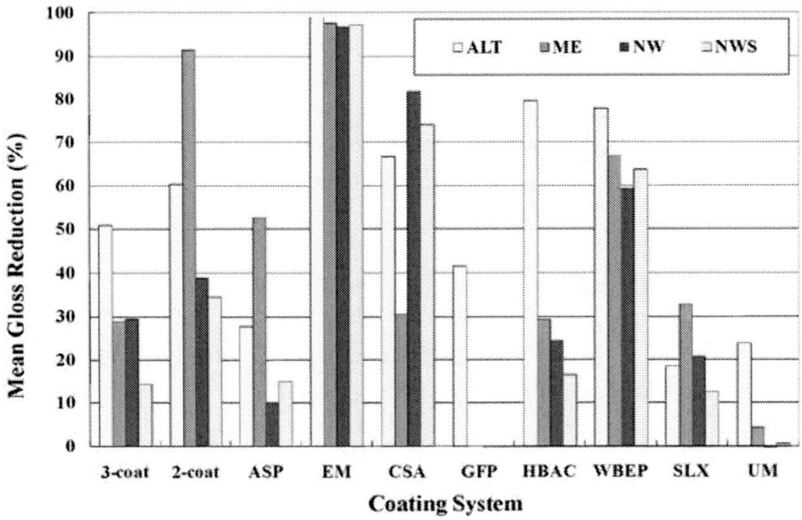

Figure 13. Graph. Mean gloss reduction data.

The initial measurements before exposure exhibited a broad range of gloss values. GFP was a flat coating with a low gloss of 4.6; EM, SLX, and UM were highly glossy (78–93), and the rest of the coatings were semi-glossy (12–60).

ALT

All coatings demonstrated gloss reduction at varying levels after ALT. SLX, UM, and ASP had a gloss reduction less than 30 percent. SLX had the least gloss reduction of 19 percent. The strong oxidation resistance of silicon resin gave this organic-inorganic hybrid coating excellent UV radiation resistance, resulting in high-quality gloss retention properties. The AR/AL of these three coatings was very low (zero for ASP and 1.5 for SLX and UM). Good gloss retention of UM and ASP can be attributed to their AL nature.

Outdoor Exposure Testing

After 24 months of exposure in ME, 18 months of exposure in NW, and 18 months of exposure in NWS, all coating systems displayed gloss reduction. UM had the least gloss reduction of 4 percent in ME. The same coating system had almost zero gloss reduction in NW and NWS. SLX, HBAC, and the three-coat system had about 30 percent gloss reduction in ME. ASP had less than 20 percent gloss retention in NW and NWS; however, it had about 50 percent gloss reduction in ME. The two-coat system that had the ASP top coat had similar behavior as that of the one-coat ASP. The two-coat system had relatively low gloss reduction of less than 39 percent in NW and NWS but had 91 percent gloss reduction in ME. HRCSA, on the other hand, had 31 percent gloss reduction in ME, which was much lower than the 74 and 82 percent reduction obtained in NW and NWS.

In summary, UM, SLX, the three-coat system, and HBAC performed best in terms of gloss retention in outdoor exposures. ASP and the two-coat system had a large loss of gloss in ME exposure. HRCSA had a large gloss reduction in NW and NWS. EM and WBEP had large gloss reductions in all test environments.

Change of Color

In addition to gloss, color is an important parameter in evaluating the weatherability of coating systems. Table 11 summarizes ΔE, and figure 14 shows the corresponding graph.

Table 11. Summary of ΔE data

Coating System	Exposure Condition			
	ALT	ME	NW	NWS
Three-coat	1.2	1.0	1.0	1.0
Two-coat	1.4	3.5	0.5	0.3
ASP	1.4	1.6	0.3	0.4
EM	8.6	9.6	14.4	15.3
HRCSA	6.3	9.8	6.3	8.2
GFP	8.2			
HBAC	10.9	2.2	3.3	3.3
WBEP	4.7	1.9	1.5	1.7
SLX	3.1	0.4	0.8	0.4
UM	3.7	0.4	0.2	0.2

Note: The blank cells indicate that no outdoor exposure data were available.

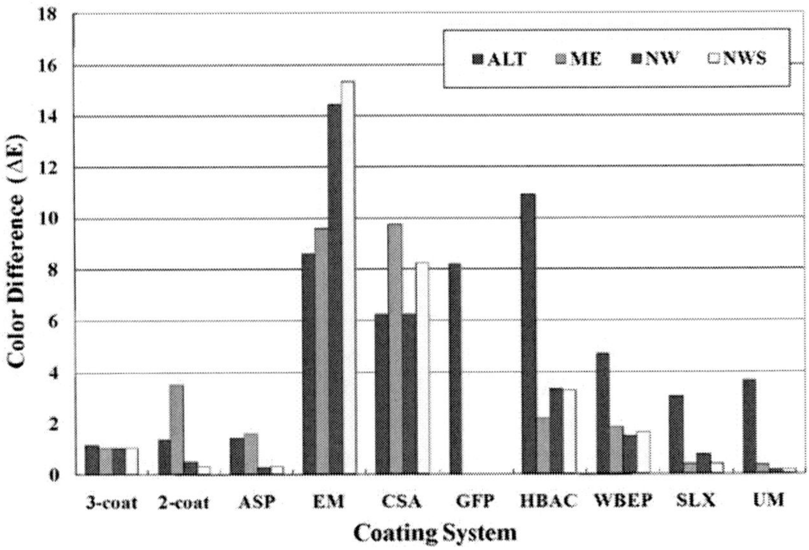

Figure 14. Graph. ΔE after exposure tests.

ALT

After ALT, the three-coat and two-coat systems had the best color retention, as indicated by low ΔE values of 1.2 and 1.4, respectively. The color retention of the tested panels compared to the nontested panels of these two coating systems was clearly noticeable by visual examination. Although ASP had a small ΔE of 1.4, some white color stains were observed on the panel surface at end of the ALT. SLX and UM also had low ΔE values of 3.1 and 3.7, respectively, indicating good color retention. On the other hand, HBAC, GFP, and EM had large ΔE values between 8.2 and 10.9. EM showed noticeable color change after the first test cycle. Panel surfaces gradually changed color from a shiny grey to a yellowish dark green. WBEP and HRCSA had moderate ΔE of 4.7 and 6.3, respectively.

Outdoor Exposure Testing

Most coating systems had similar color retention characteristics in the outdoor exposures. However, the two-coat system and ASP had good color retention in NW and NWS but had a noticable color change after exposure in ME. Visual observation revealed that the color on the overall surface of the panels faded in ME.

Most coating systems exhibited good color retention except for HRCSA and EM. The ΔE of HBAC after outdoor exposure was much smaller compared to the changes after ALT. As mentioned earlier, due to the softness and stickiness of HBAC, the surface of the test panels picked up some dirt, so the panel surfaces looked dirty and dark after ALT. This phenomenon was not observed in outdoor exposures. The surface darkness of laboratory tested panels may have affected the color and gloss value.

Table 12. Pencil scratch hardness data

Coating System	Initial	Final			
		ALT	ME	NW	NWS
Three-coat	HB	HB	HB	HB	HB
Two-coat	HB	HB	2H	2H	2H
ASP	6B	4B	4B	4B	4B
EM	HB	HB	HB	HB	HB
HRCSA	<6B	<6B	<6B	<6B	<6B
GFP	2H	2H			
HBAC	<6B	<6B	<6B	<6B	<6B
WBEP	HB	HB	HB	HB	HB
SLX	HB	2H	2H	2H	2H
UM	2B	HB	HB	HB	HB

Note: The blank cells indicate that no outdoor exposure data were available.

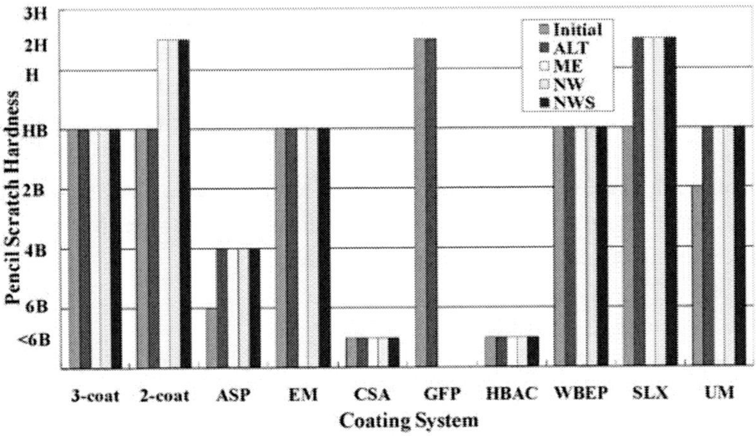

Figure 15. Graph. Pencil scratch hardness before and after exposure tests.

Change of Pencil Scratch Hardness

Pencil scratch hardness data before and after the exposure tests are summarized in table 12. Figure 15 shows the bar graph of pencil scratch hardness data. In this table, "H" represents hardness, "B" represents blackness, and "HB" represents hard and black pencils. The different grades of hardness are as follows:

9H (hardest) > 8H > 7H > 6H > 5H > 4H > 3H > 2H > H > HB > B > 2B > 3B > 4B > 5B > 6B > 7B > 8B > 9B (softest)

ALT

GFP was the hardest coating system, with an initial pencil scratch hardness of 2H. EM, WBEP, and SLX had initial pencil scratch hardness of HB, similar to that of the two controls. UM and ASP, with scratch hardness of 2B and 6B, respectively, were relatively softer. HRCSA and HBAC were the softest coating systems, as the top coat of these two coatings scratched easily when a 6B pencil was applied. Several damaged areas were created on the surface of a few HRCSA test panels even after careful handling. Both HRCSA and HBAC were very soft and sticky. As a result, they exhibited the tendency to collect dust and external airborne material that resulted in an unclean appearance. Most of the coating systems had no hardness changes after ALT except for ASP, SLX, and UM, which became harder.

Outdoor Exposure Testing

The two-coat system, ASP, SLX, and UM had the same degree of hardness increase after the three outdoor exposures. Hardness of the other systems remained unchanged after the outdoor exposures.

Adhesion Strength

Table 13 lists the initial adhesion strength of the 10 coating systems obtained with a hydraulic tester and a pneumatic tester. Figure 16 shows the comparison of test results by these two methods. Both methods resulted in similar data for all coating systems except for the two-coat system and GFP with CVs of 28 and 22 percent, respectively. The three-coat system, EM, WBEP, SLX, and UM had initial adhesion strength greater than 1,500 psi (10,335 kPa), while the two-coat system, ASP, HRCSA, GFP, and HBAC had initial adhesion strength lower than 1,000 psi (6,890 kPa). HRCSA exhibited the weakest adhesion strength of 366 psi (2,521.74 kPa).

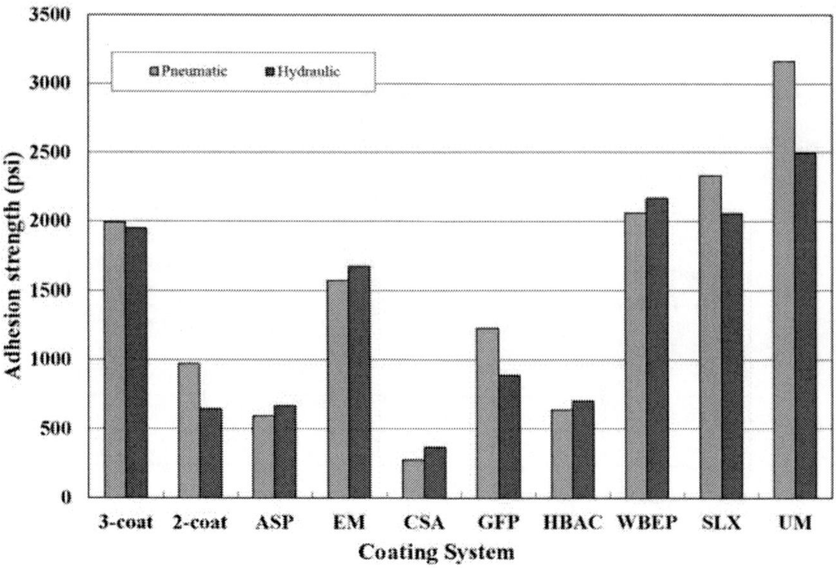

1 psi = 6.89 kPa.

Figure 16. Graph. Comparison of initial adhesion strength data using pneumatic and hydraulic methods.

Table 13. Initial adhesion strength from hydraulic and pneumatic test methods

Coating System	Pneumatic Tester (psi)	Hydraulic Tester (psi)	Mean (psi)	Standard Deviation (psi)	CV (percent)
Three-coat	1,192	1,948	1,970	31.3	1.6
Two-coat	968	646	807	227.8	28.2
ASP	592	664	628	51.3	8.2
EM	1,1571	1,676	1,624	73.9	4.6
HRCSA	276	366	321	63.8	19.9
GFP	1,226	886	1,056	240.3	22.8
HBAC	635	700	668	45.6	6.8
WBEP	2,061	2,168	2,114	76.0	3.6
SLX	2,332	2,057	2,194	194.7	8.9
UM	3,160	2,492	2,826	472.5	16.7

1 psi = 6.89 kPa.

Table 14. Mean adhesion strength changes after ALT and outdoor exposure tests

Coating System	Exposure Condition (percent)							
	ALT		ME		NW		NWS	
	Unscribed	Scribed	Unscribed	Scribed	Unscribed	Scribed	Unscribed	Scribed
Three-coat	-12	-37	-3	-4	3	-10	18	5
Two-coat	107	60	10	10	46	35	23	31
ASP	4		-8	-6	34	39	20	23
EM	-23	-34	-17	-2	3	5	-2	-3
HRCSA	13	11	17	9	20	16	29	24
GFP	30	20						
HBAC	-12	75	8	53	85	25	36	87
WBEP	-32	-46	-2	1	-21	-3	-18	8
SLX	-22	-44	-18	-50	4	-17	-15	3
UM	-22		-10	-19	-5	-11	-6	7

Note: The blank cells indicate that no outdoor exposure data were available.

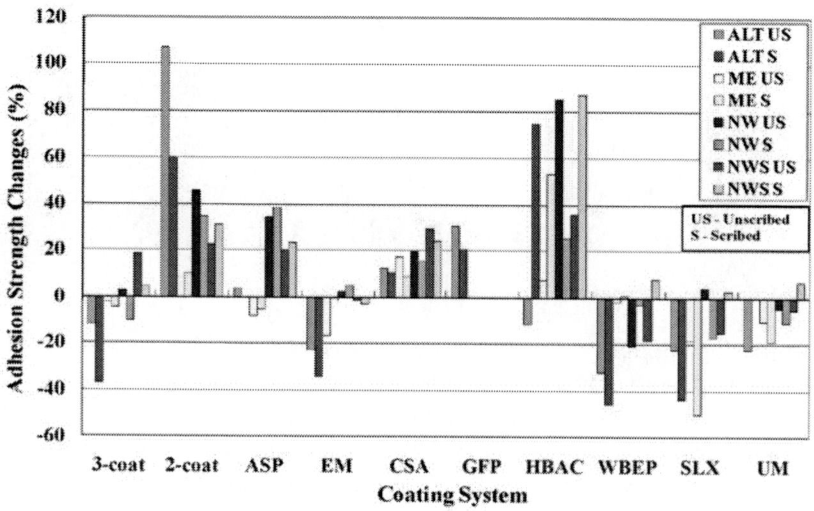

Figure 17. Graph. Changes in mean adhesion strength after ALT and outdoor tests.

Table 14 summarizes the mean adhesion strength changes of the test panels after the accelerated and outdoor exposures, while figure 17 shows the plot of these adhesion strength changes.

ALT

Adhesion strength data for scribed test panels of ASP and UM could not be obtained after ALT because the panel surface had too many blisters and other surface defects to find a flat surface to glue the dollies.

The three-coat system demonstrated a cohesive failure mode with failure occurring within the epoxy zinc-rich primer, while the two-coat system exhibited an adhesive failure mode at the interface between the zinc-rich primer and the ASP top coat. All one-coat systems had a cohesive failure mode except for SLX, which had a partially cohesive failure and partially adhesive failure between the coating and the substrate.

The three-coat system, EM, and WBEP had a shallow cohesive failure mode where residual DFT were close to their initial DFT, and the failure mode was observed close to the panel surface. Figure 18 shows the cohesive failure mode of the ASP and EM test panels. ASP had an initial DFT of 20 mil (508 μm), leaving 4 mil (101.6 μm) of DFT on the adhesion spot after testing. EM coating system had an initial DFT of 9 mil (228.6 μm) and had 8 mil (203.2 μm) of DFT remaining on the pull-off spot after the test.

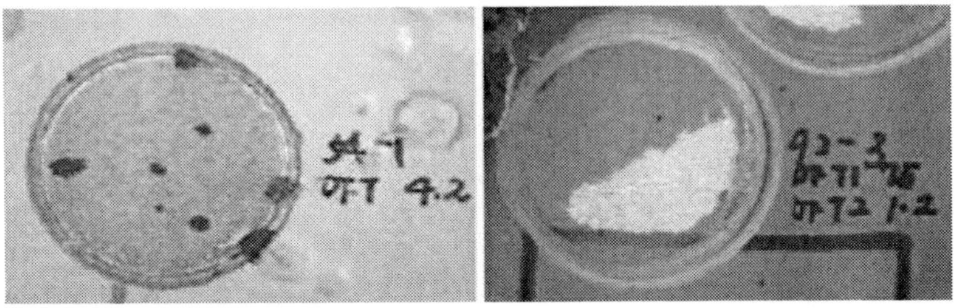

Figure 18. Photo. Cohesive failure modes of ASP and EM.

Adhesion strengths varied with either an increase or decrease after ALT. ASP had the lowest adhesion strength increase (4 percent), while the two-coat system had the highest adhesion strength increase (>100 percent) among unscribed panels. The three-coat system (12 percent) and WBEP (32 percent) had the lowest and highest decrease in adhesion strength variation among unscribed panels. HRCSA (11 percent) and HBAC (75 percent) had the lowest and highest increase in adhesion strength, while EM (34 percent) and WBEP (46 percent) had the lowest and highest decrease in adhesion strength among scribed panels. CV of the adhesion test data ranged from 3 to 36 percent with a median value of 16 percent.

Outdoor Exposure Testing

Most coating systems did not show significant adhesion strength changes but had cohesive failure modes after outdoor exposures. The three-coat system failed cohesively within the primer. The two-coat system exhibited an adhesion failure mode between the top coat and the primer.

Scribed SLX test panels lost 50 percent adhesion strength after 24 months in ME due to severe rust creepage. HBAC had an obvious increase in adhesion strength after all three types of outdoor testing. This was probably due to additional long-term curing of the resin in the exposure environments.

Surface Defects, Rusting, and Blistering

Representative progressive changes of test surface condition with time of coating systems tested in this study are shown in figure 19 through figure 55. Table 15 summarizes the surface blisters, rusting, and defects developed in the laboratory and outdoor exposures. Figure 56 through figure 59 show the cumulative number of surface defects identified by the holiday detector for each coating system after ALT, ME, NW, and NWS, respectively. The acronyms used in table 15 are based on ASTM D714-02, “Standard Test Method for Evaluating Degree of Blistering of Paints” and ASTM D610-08,

"Standard Test Method for Evaluating Degree of Rusting on Painted Steel Surfaces."[31,32]

The blister and rusting abbreviations have been specified in these standards, which have been used to categorize the coating surface degradation. In this grading, "F" denotes few, "M" denotes medium, and "D" denotes dense. Size 2 is the largest while size 8 is the smallest.

The blister grading is as follows:

F8 = Few blister size 8, 4M = Medium blister size 4, 2 MD = Medium dense blister size 2, F6 = Few blister size 6, F2 = Few blister size 2, 4 MD = Medium dense blister size 4, and 2D = Dense blister size 2.

The rusting grades have been progressively assigned 9 through 1. G9, G8, G6, G5, G4, and G1 denote rusting grades 9, 8, 6, 5, 4, and 1, respectively. G9 covers less than 0.03 percent of the area, while G2 covers half of the panel.

ALT

The three-coat system (see figure 19) and GFP (see figure 39) retained the best surface physical properties after completion of ALT. One rust pit was observed on one of the GFP panels. Additionally, one rusted blister (size 8) was observed, and one defect was detected in one of the three-coat test panels after 6,840 h of testing.

HRCSA (see figure 35) developed a few size 6 blisters on several test panels, and two holidays were detected after 2,160 and 4,320 h, respectively. Because HRCSA was soft in surface nature, there were a few damaged spots on almost every HRCSA test panel, although they were handled carefully. The panel surfaces looked dirty because they picked up a lot of dust and dirt due to the soft and sticky nature of the system.

HBAC, WBEP, and SLX had moderate surface failures. HBAC developed one size 6 rusted blister on panel surface. A few defects were detected by the holiday detector during the test period. Although this system was soft, there were no damaged spots. These panels also looked dirty due to the stickiness of the coating system. WBEP developed a few size 6 and size 2 blisters, as well as a few defects.

Table 15. Development of blistering, rusting, and surface defects in ALT and outdoor exposure tests

Coating System	ALT			ME Exposure			Mild NW Exposure			Mild NWS Exposure			Overall Defects
	Blister	Rusting	Defect	Blister	Rusting	Defect	Blister	Rusting	Defect	Blister	Rusting	Defect	
Three-coat	F8 (1P)	G9	1	0	0	1	0	0	1	0	0	0	3
Two-coat	0	0	>100	0	0	0	0	0	1	0	0	0	>100
ASP	4M, 2MD (all)	0	>100	0	G9	0	0	0	4	0	0	0	>100
EM	0	0	>100	0	0	35	0	0	12	0	0	12	>100
HRCSA	F6 (3P)	0	2	F8, F6 (3P)	0	0	0	0	1	0	0	0	3
GFP	0	1	0										0
HBAC	F6 (IP)	G9	7	0	0	0	0	0	1	0	0	0	8
WBEP	F6 (2P), F2 (1P)	0	5	F6 (3P)	G9	13	F6 (4P)	0	39	F6 (all)	G6	>100	>100
SLX	0	G9	32	F2(1P)	G9	11	0	0	68	0	G9	28	>100
UM	4MD,2D (all)	G4, G1	>100	F8, M8 (all)	G8, G5	>100	0	0	67	0	0	30	>100

1P = One panel; 2P = Two panels; 3P = Three panels; and all = All panels.

Note: The blank cells indicate that no outdoor exposure data were available for GFP.

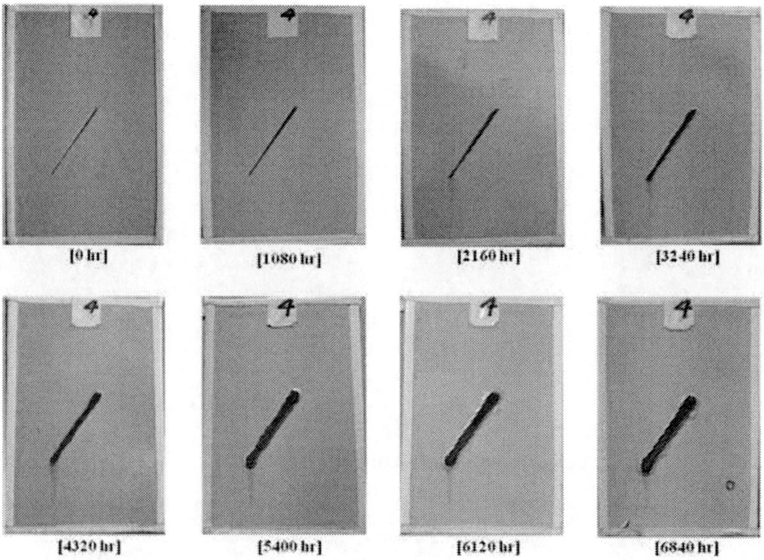

Figure 19. Photo. Progressive changes of panel 4 (three-coat: ALT).

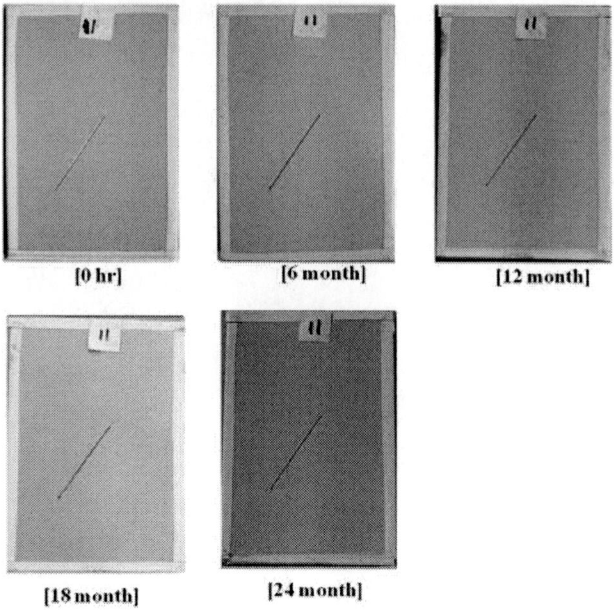

Figure 20. Photo. Progressive changes of panel 11 (three-coat: ME).

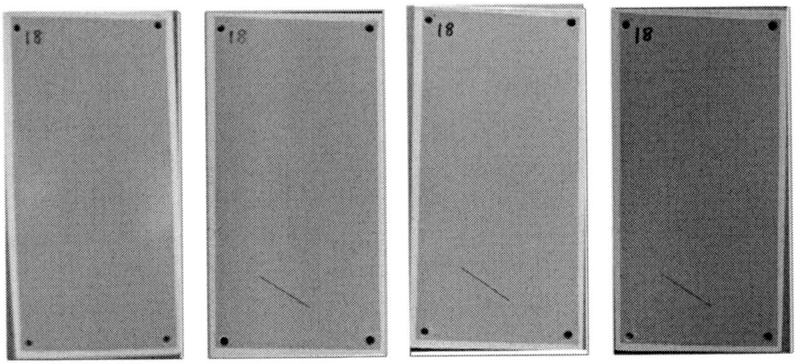

[0 hr] [7 month] [13 month] [18 month]

Figure 21. Photo. Progressive changes of panel 18 (three-coat: NW).

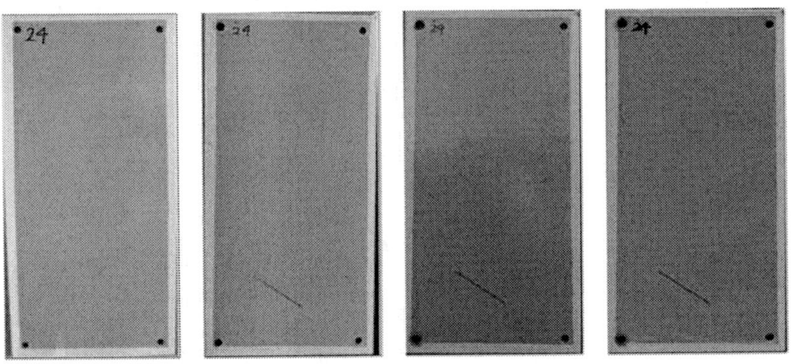

[0 hr] [7 month] [13 month] [18 month]

Figure 22. Photo. Progressive changes of panel 24 (three-coat: NWS).

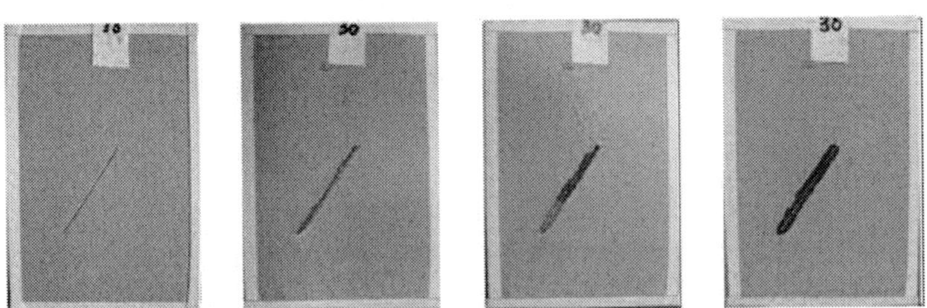

[0 hr] [1080 hr] [2160 hr] [3240 hr]

Figure 23. (Continued).

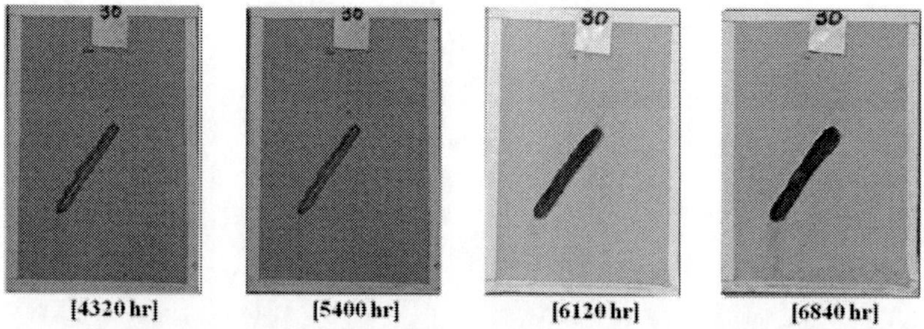

Figure 23. Photo. Progressive changes of panel 30 (two-coat: ALT).

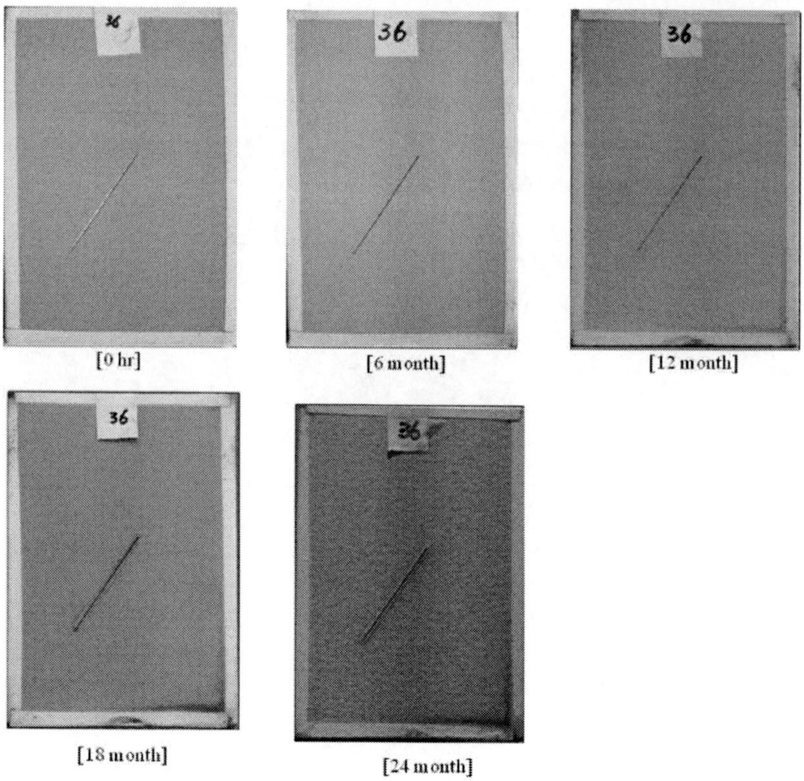

Figure 24. Photo. Progressive changes of panel 36 (two-coat: ME).

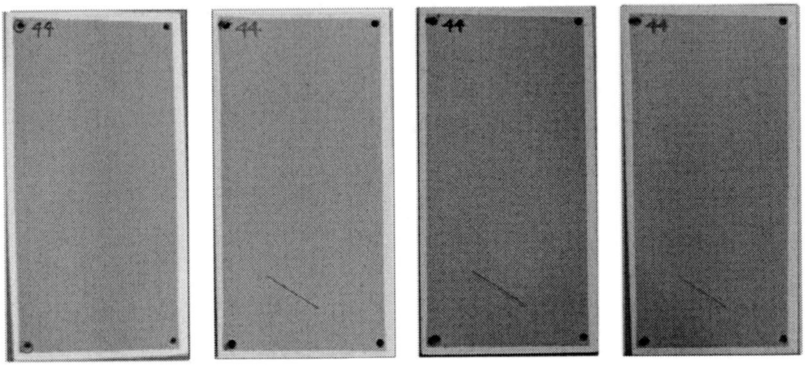

[0 hr] [6 month] [13 month] [18 month]

Figure 25. Photo. Progressive changes of panel 44 (two-coat: NW).

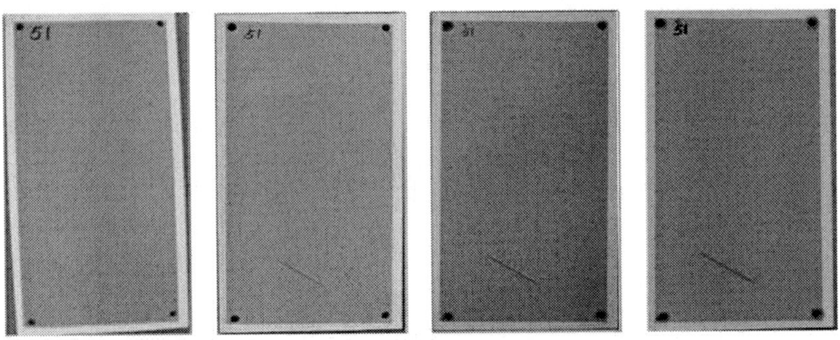

[0 hr] [7 month] [13 month] [18 month]

Figure 26. Photo. Progressive changes of panel 51 (two-coat: NWS).

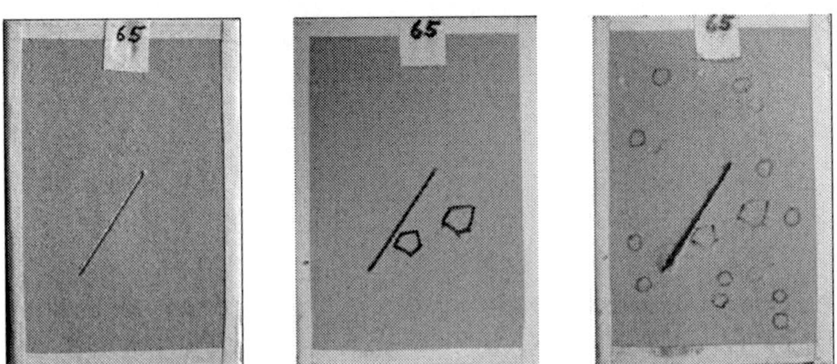

[0 hr] [1080 hr] [2160 hr]

Figure 27. (Continued).

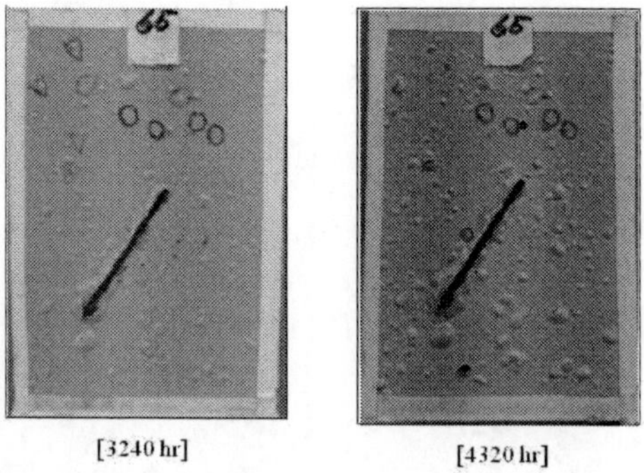

Figure 27. Photo. Progressive changes of panel 65 (ASP: ALT).

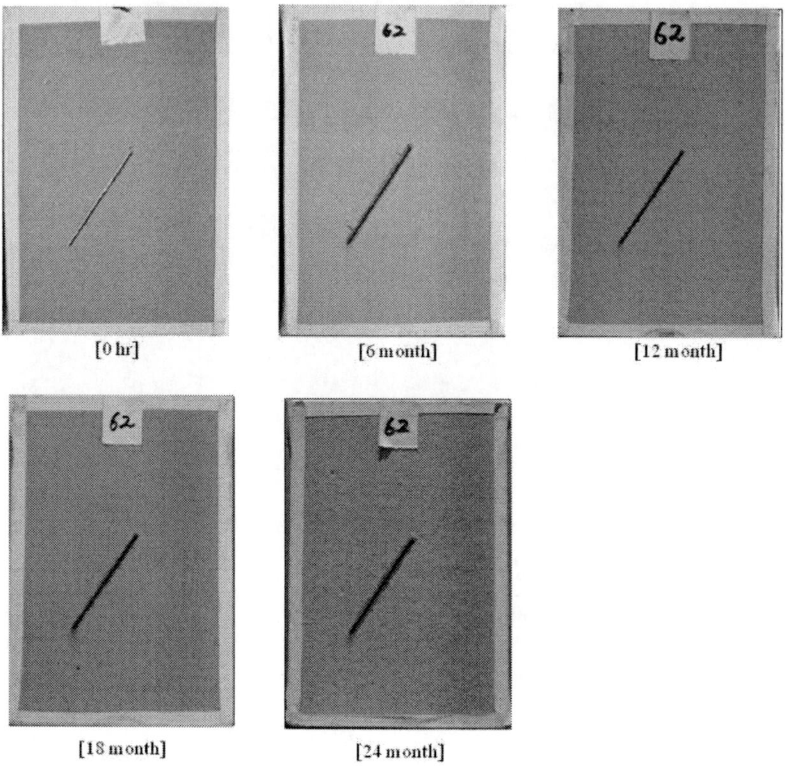

Figure 28. Photo. Progressive changes of panel 62 (ASP: ME).

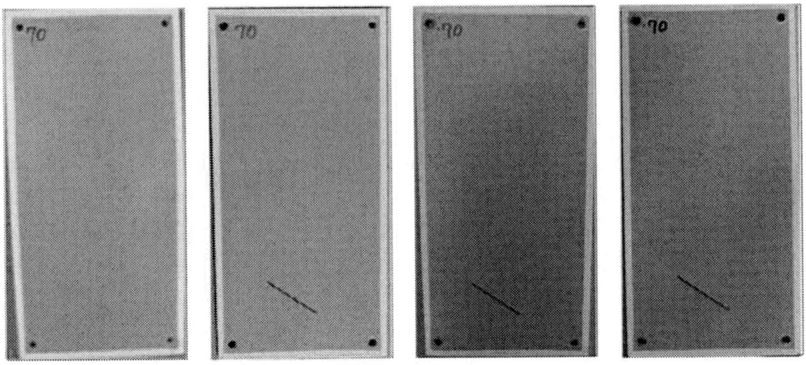

[0 hr] [7 month] [13 month] [18 month]

Figure 29. Photo. Progressive changes of panel 70 (ASP: NW).

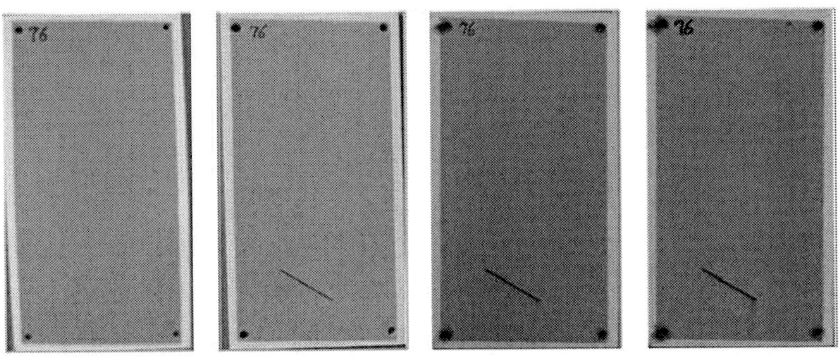

[0 hr] [7month] [13 month] [18 month]

Figure 30. Photo. Progressive changes of panel 76 (ASP: NWS).

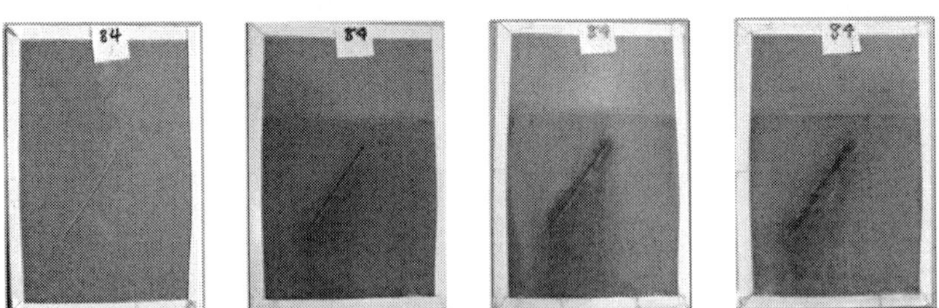

[0 hr] [1080 hr] [2160 hr] [3240 hr]

Figure 31. (Continued).

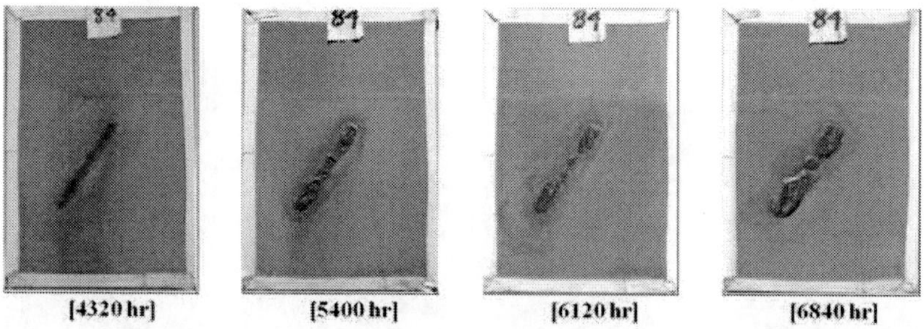

Figure 31. Photo. Progressive changes of panel 84 (EM: ALT).

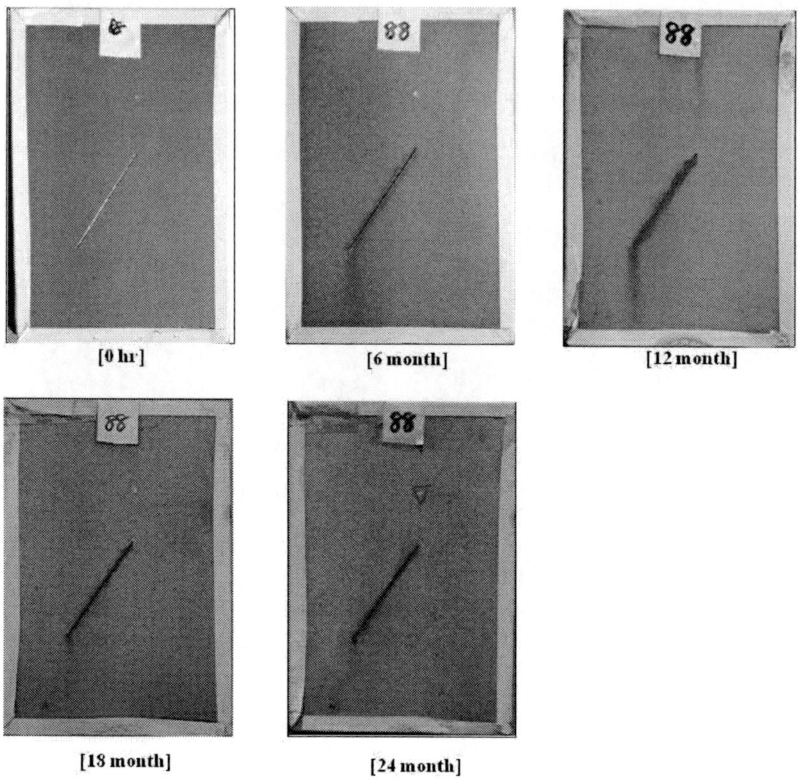

Figure 32. Photo. Progressive changes of panel 88 (EM: ME).

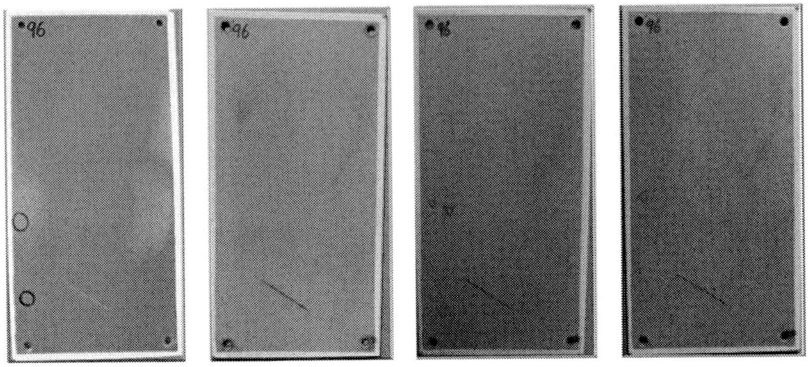

[0 hr] [7 month] [13 month] [18 month]

Figure 33. Photo. Progressive changes of panel 96 (EM: NW).

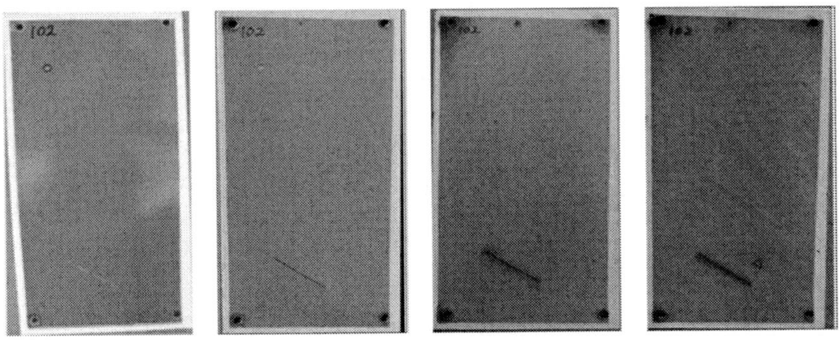

[0 hr] [7 month] [13 month] [18 month]

Figure 34. Photo. Progressive changes of panel 102 (EM: NWS).

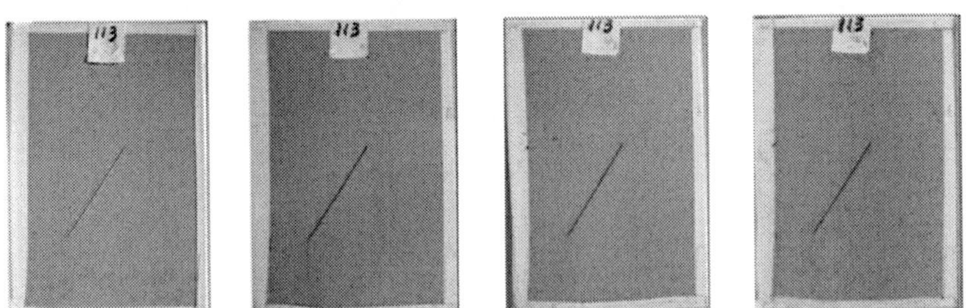

[0 hr] [1080 hr] [2160 hr] [3240 hr]

Figure 35. (Continued).

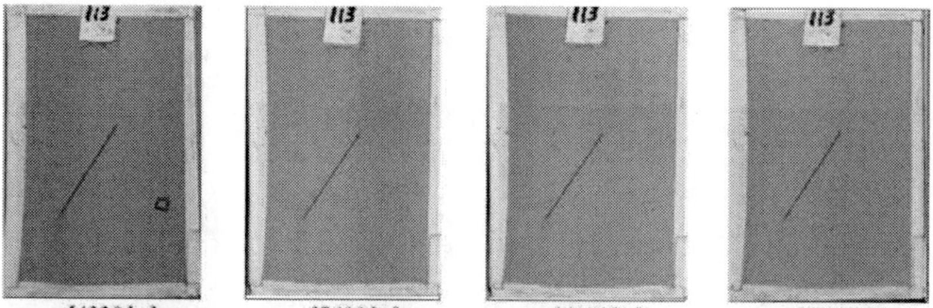

[4320 hr] [5400 hr] [6120 hr] [6840 hr]

Figure 35. Photo. Progressive changes of panel 113 (HRCSA: ALT).

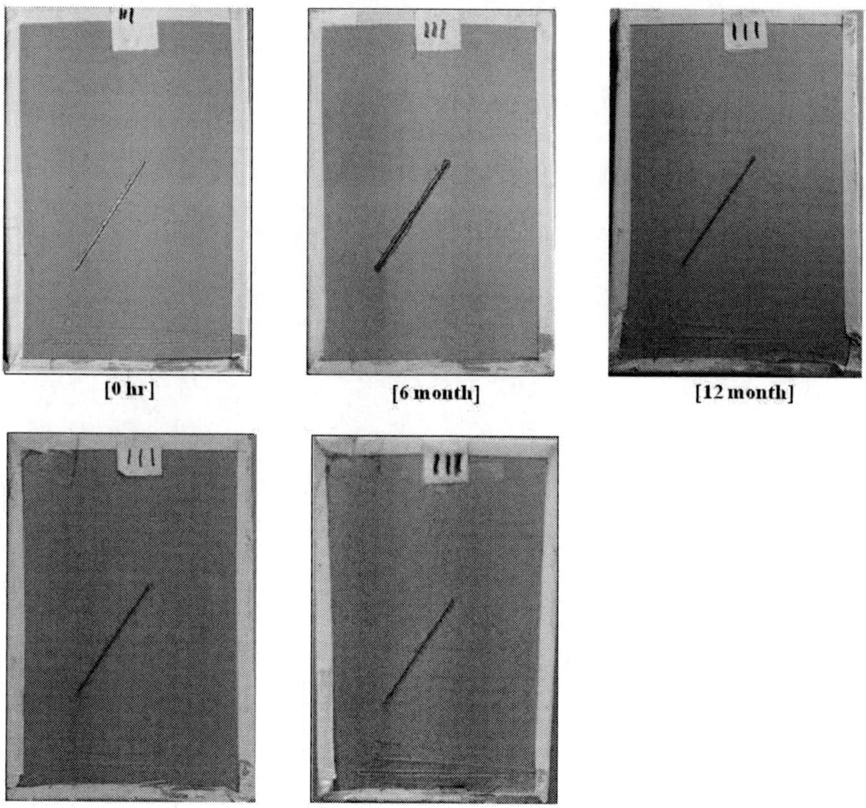

[0 hr] [6 month] [12 month]

[18 month] [24 month]

Figure 36. Photo. Progressive changes of panel 111 (HRCSA: ME).

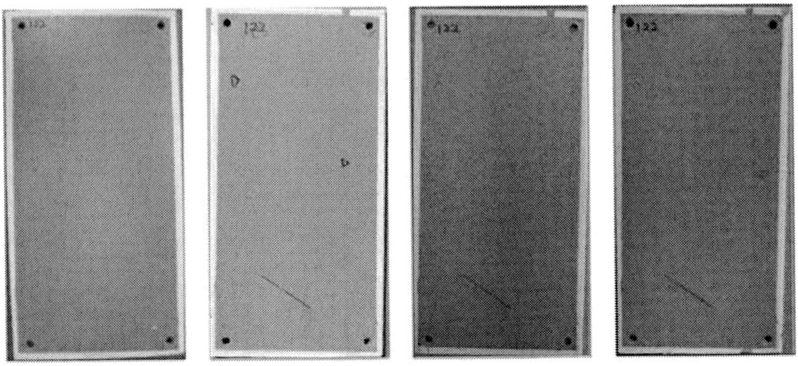

[0 hr] [7 month] [13 month] [18 month]

Figure 37. Photo. Progressive changes of panel 122 (HRCSA: NW).

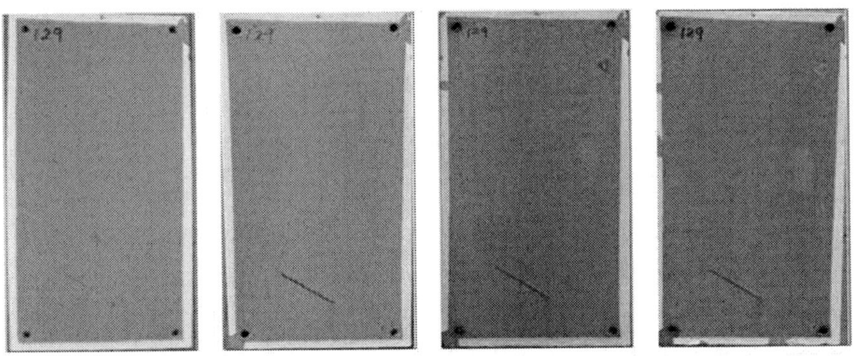

[0 hr] [7 month] [13 month] [18 month]

Figure 38. Photo. Progressive changes of panel 129 (HRCSA: NWS).

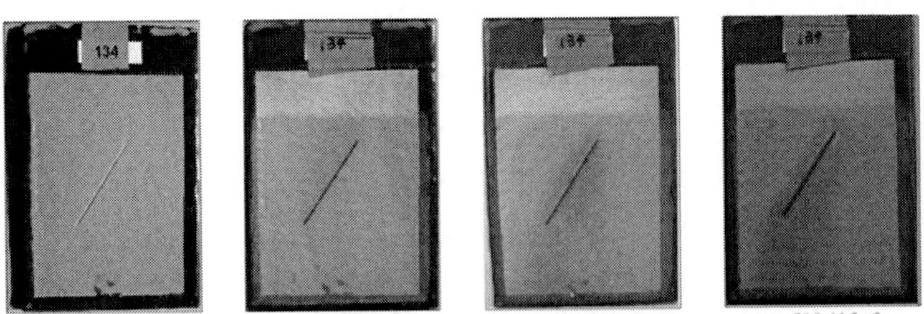

[0 hr] [1080 hr] [2160 hr] [3240 hr]

Figure 39. (Continued).

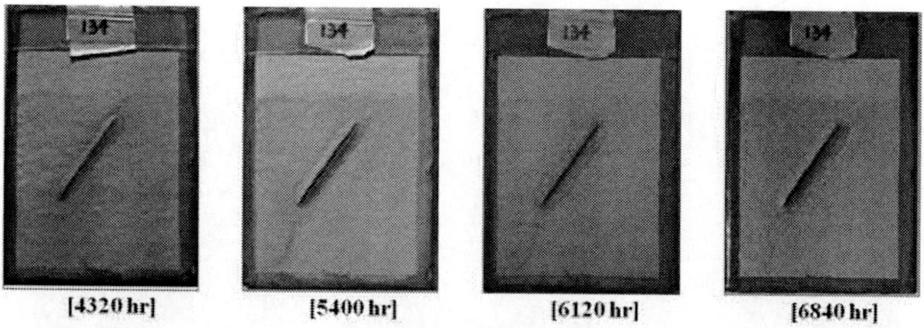

Figure 39. Photo. Progressive changes of panel 134 (GFP: ALT).

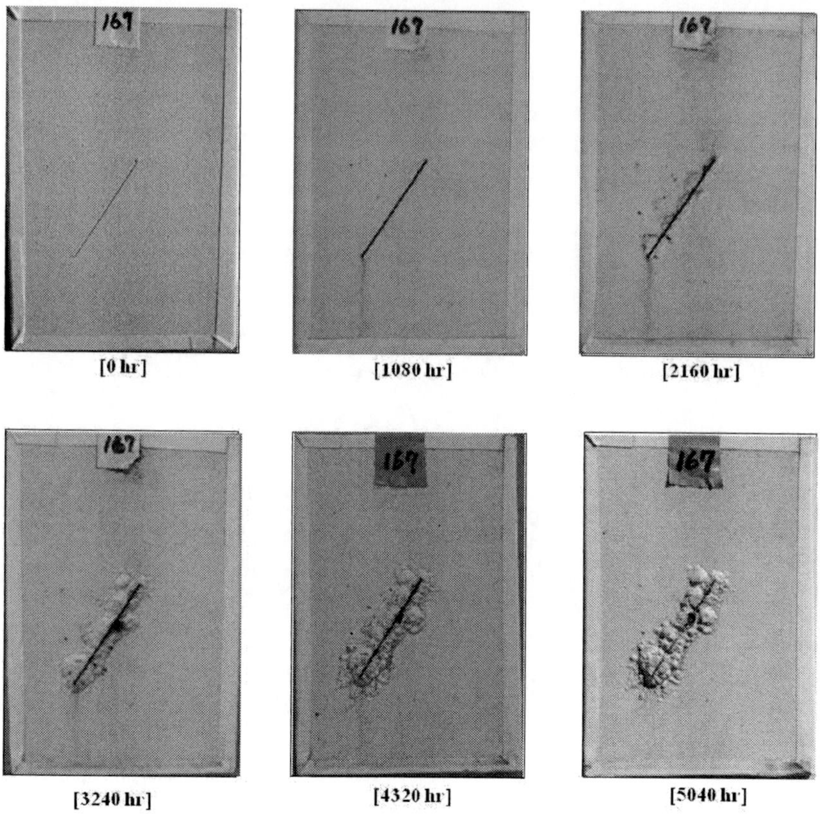

Figure 40. Photo. Progressive changes of panel 167 (HBAC: ALT).

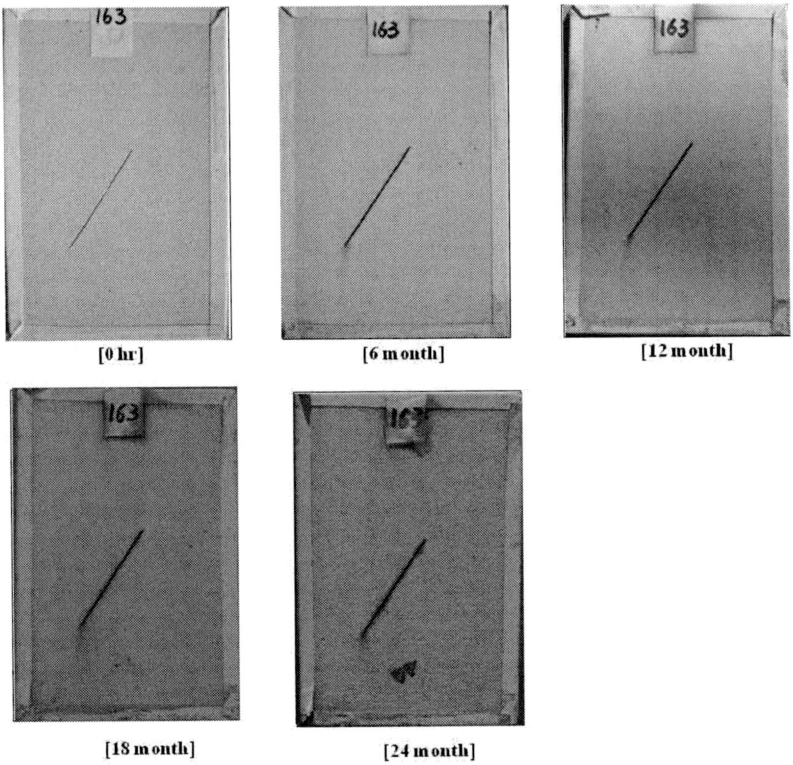

Figure 41. Photo. Progressive changes of panel 163 (HBAC: ME).

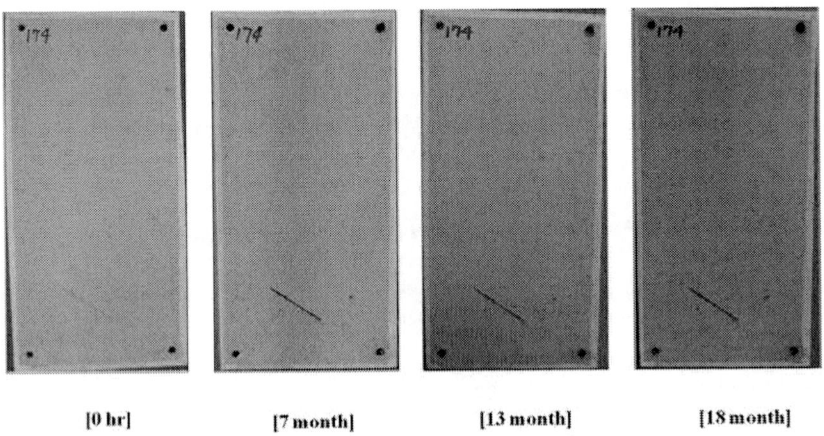

Figure 42. Photo. Progressive changes of panel 174 (HBAC: NW).

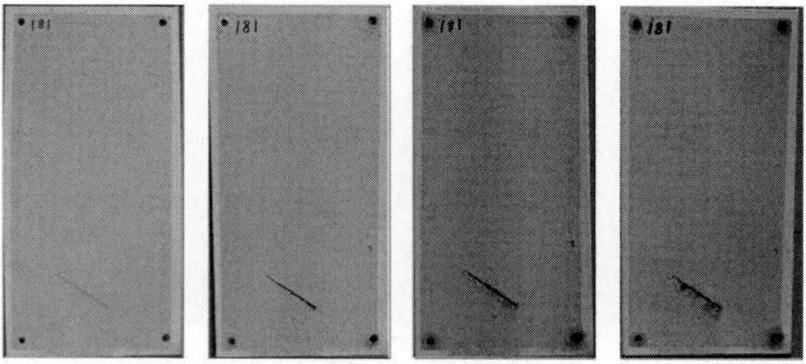

[0 hr] [7 month] [13 month] [18 month]

Figure 43. Photo. Progressive changes of panel 181 (HBAC: NWS).

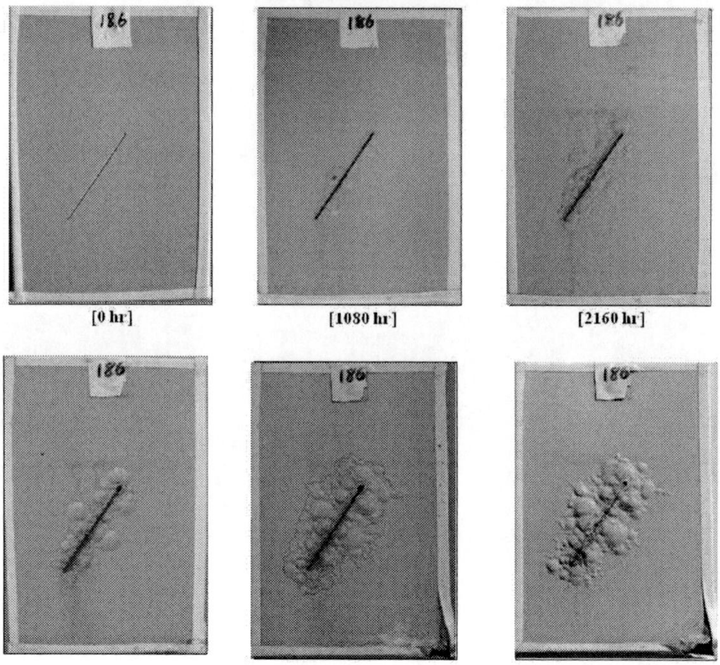

[3240 hr] [4320 hr] [5040 hr]

Figure 44. Photo. Progressive changes of panel 186 (WBEP: ALT).

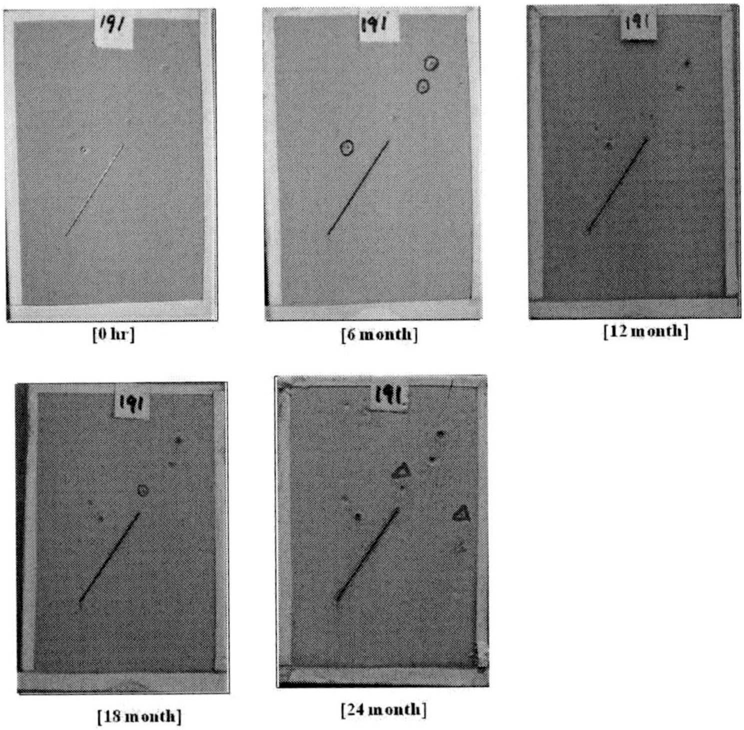

Figure 45. Photo. Progressive changes of panel 191 (WBEP: ME).

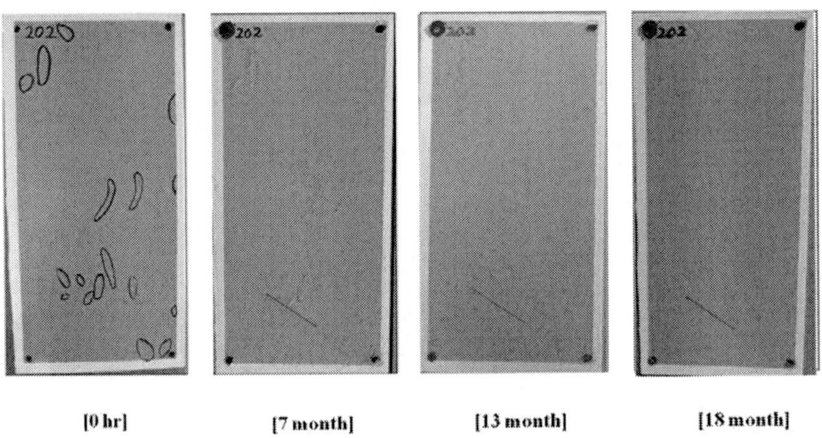

Figure 46. Photo. Progressive changes of panel 202 (WBEP: NW).

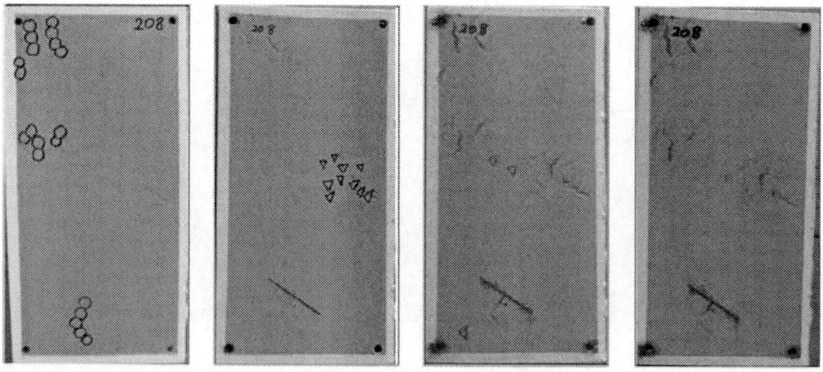

[0 hr] [7 month] [13 month] [18 month]

Figure 47. Photo. Progressive changes of panel 208 (WBEP: NWS).

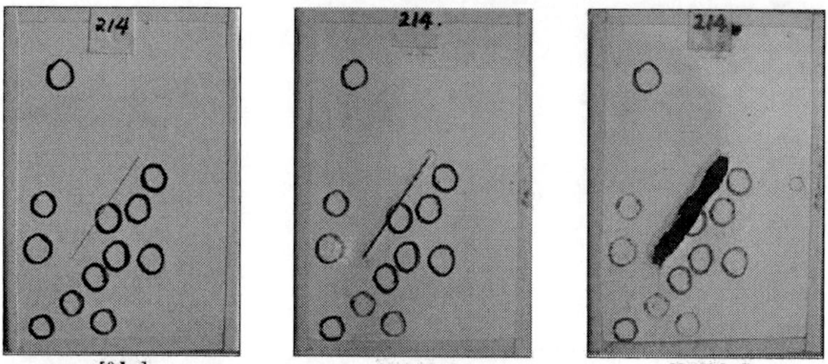

[0 hr] [1080 hr] [2160 hr]

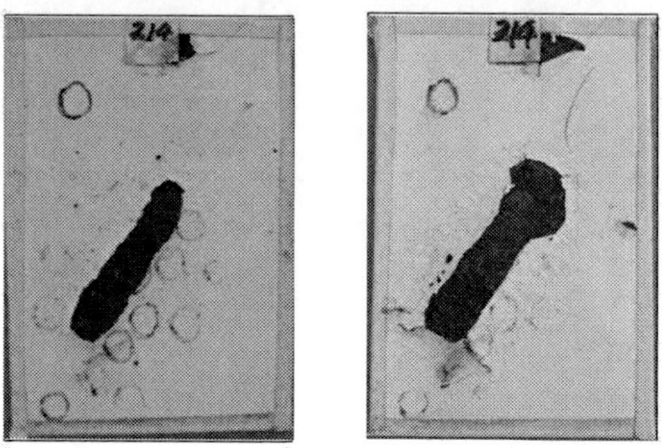

[3240 hr] [4320 hr]

Figure 48. Photo. Progressive changes of panel 214 (SLX: ALT).

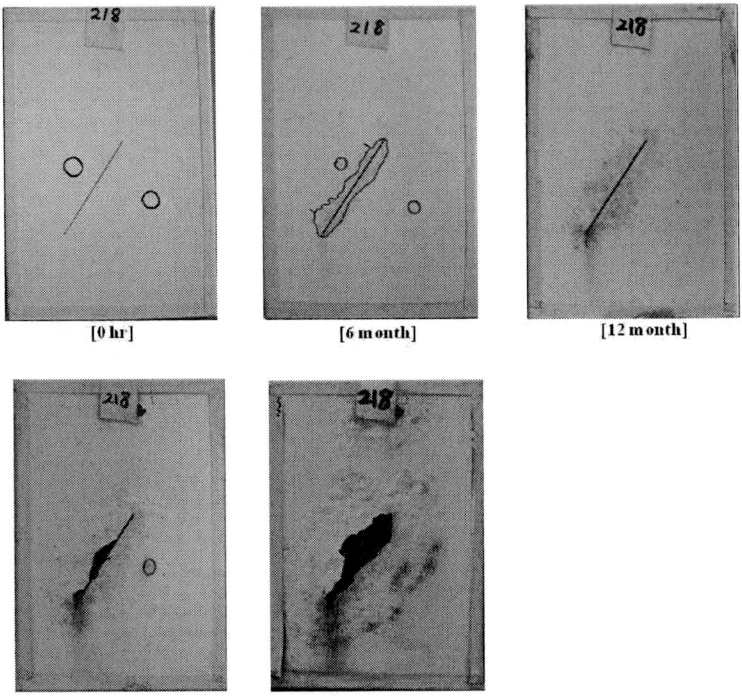

[0 hr] [6 month] [12 month]

[18 month] [24 month]

Figure 49. Photo. Progressive changes of panel 218 (SLX: ME).

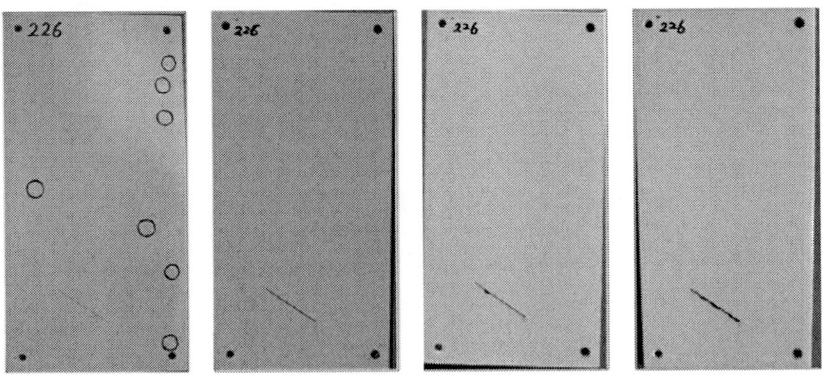

[0 hr] [7 month] [13 month] [18 month]

Figure 50. Photo. Progressive changes of panel 226 (SLX: NW).

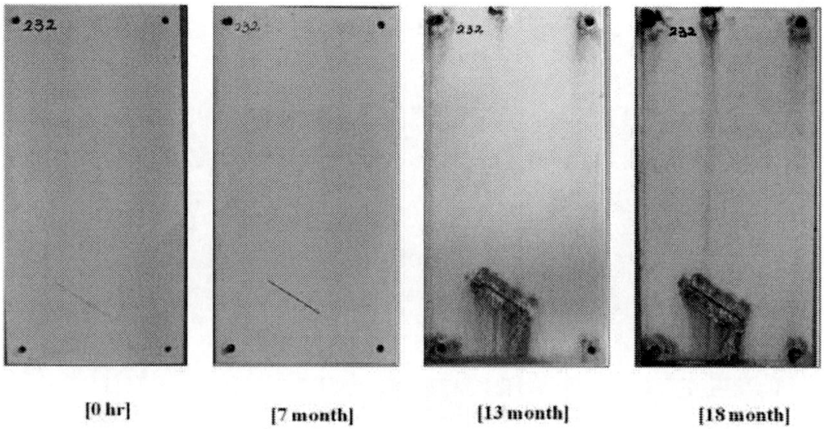

Figure 51. Photo. Progressive changes of panel 232 (SLX: NWS).

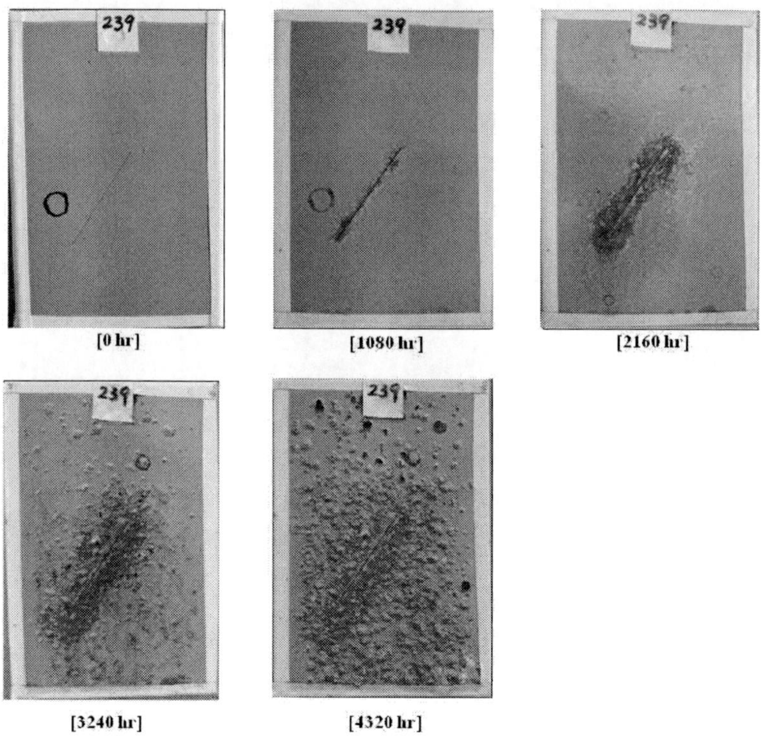

Figure 52. Photo. Progressive changes of panel 239 (UM: ALT).

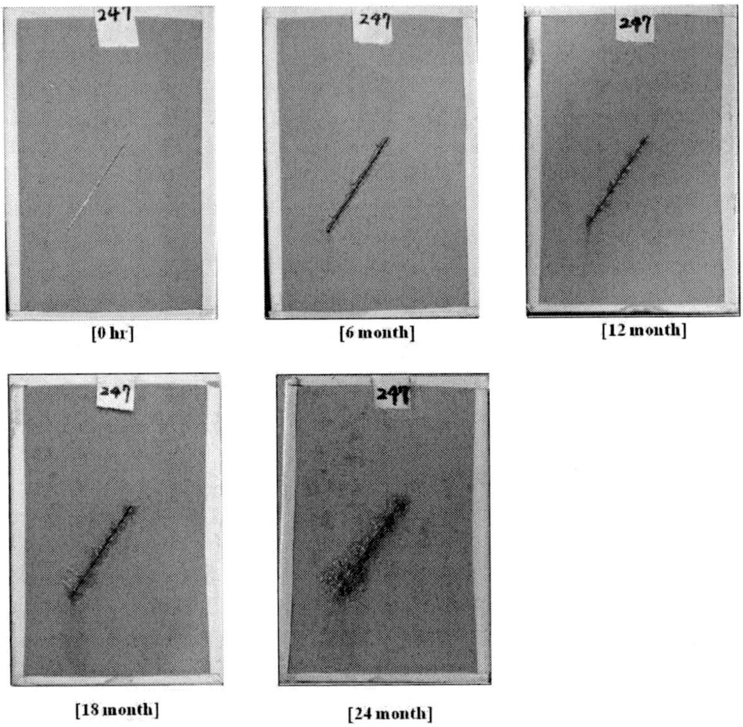

Figure 53. Photo. Progressive changes of panel 247 (UM: ME).

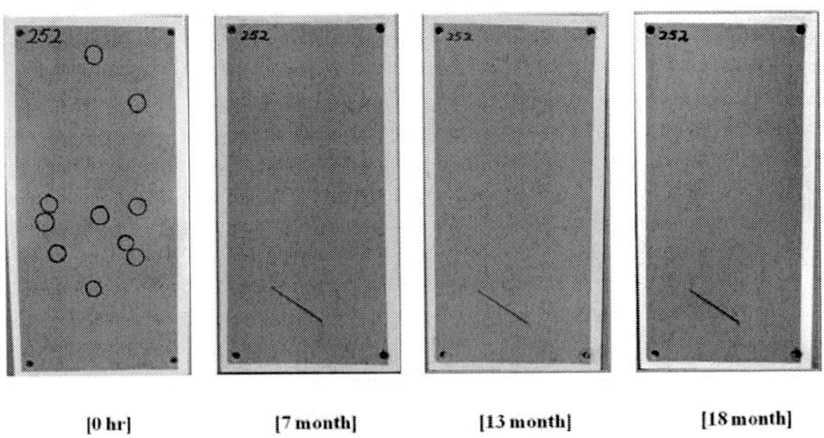

Figure 54. Photo. Progressive changes of panel 252 (UM: NW).

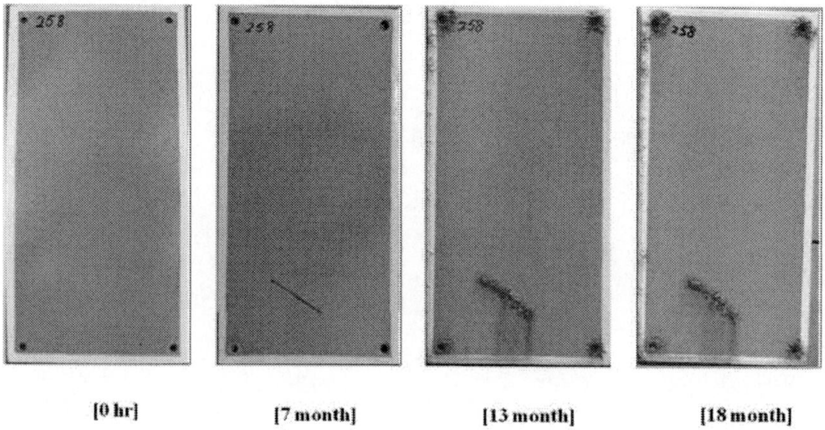

Figure 55. Photo. Progressive changes of panel 258 (UM: NWS).

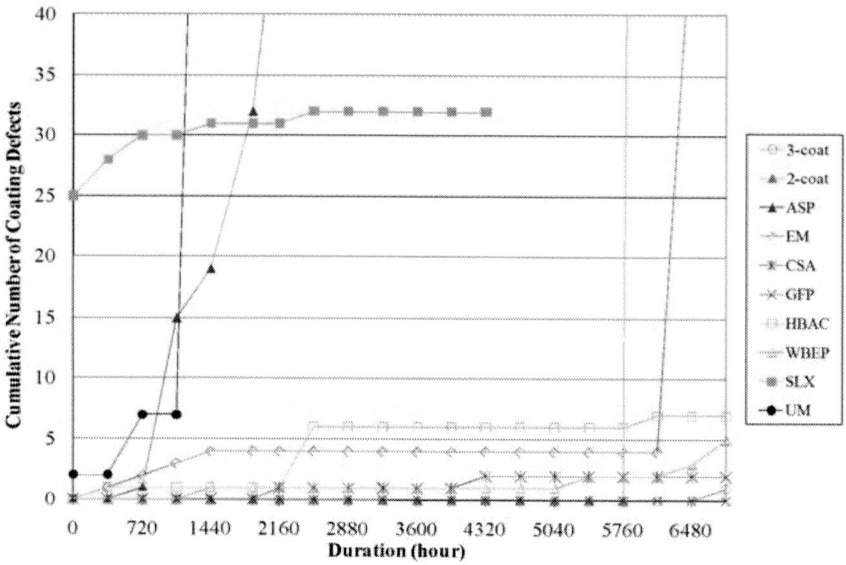

Figure 56. Graph. Development of coating defects during ALT.

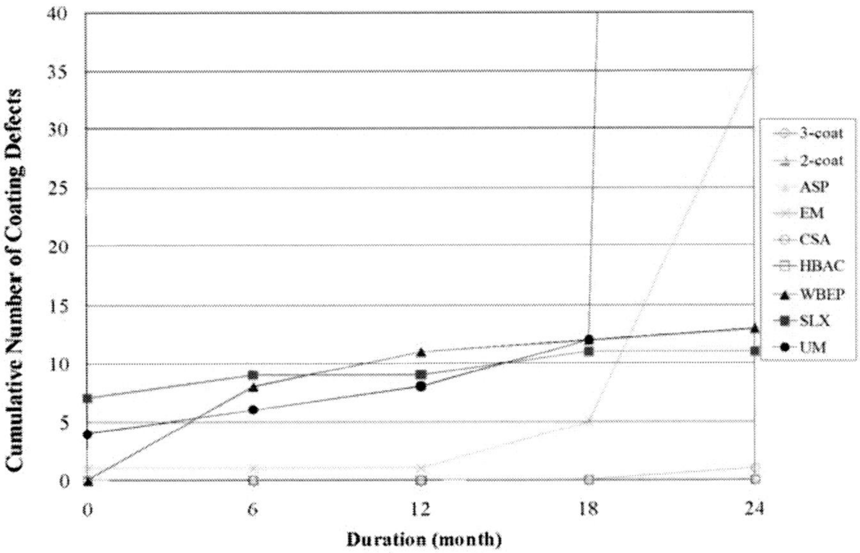

Figure 57. Graph. Development of coating defects during ME.

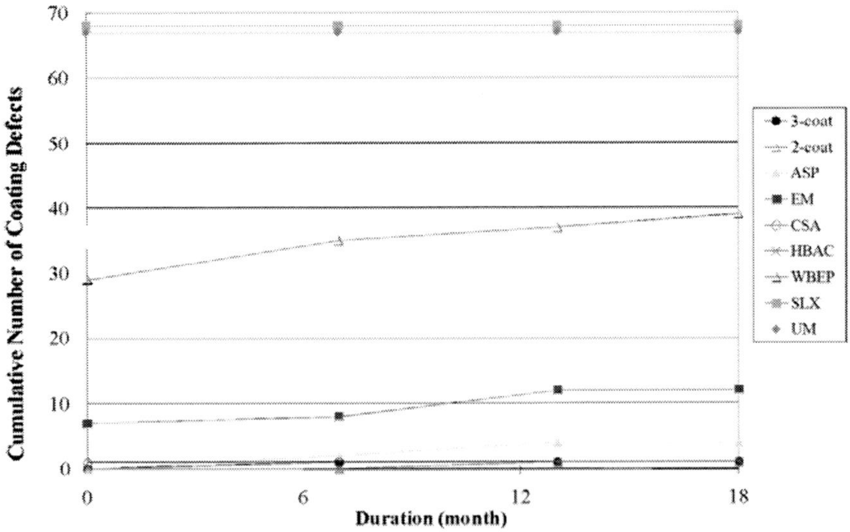

Figure 58. Graph. Development of coating defects during NW.

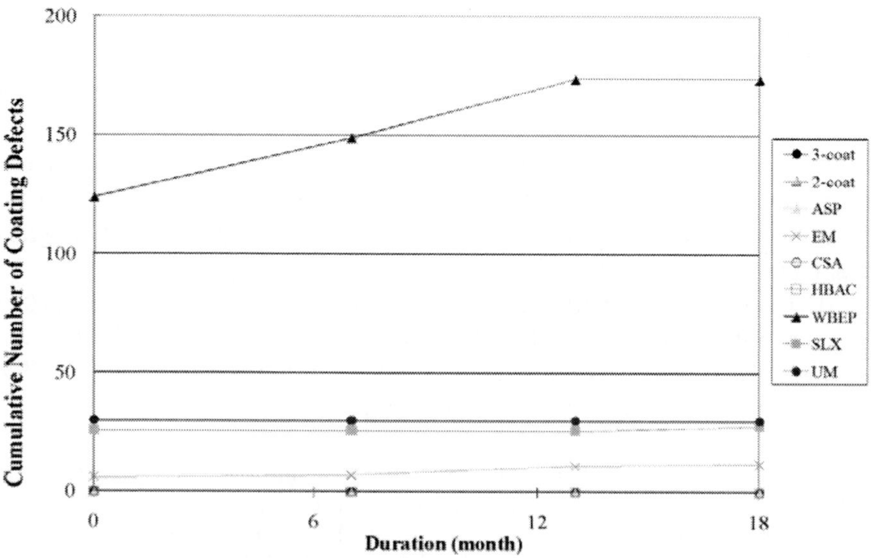

Figure 59. Graph. Development of coating defects during NWS.

There were 25 holidays on SLX test panels initially (see figure 48), and 32 defects were detected at the end of the 6,840-h test period. No blisters were observed on the panel surfaces.

ASP, UM, the two-coat control, and EM developed many surface failures. The steep lines related to these coating systems in figure 56 indicate excessive development of defects during ALT. When the excessive defects were observed, an arbitrary number of 100 was used to plot indefinable quantities.

ASP (see figure 27) exhibited one defect after 720 h of testing, and the number of defects increased rapidly. After 3,600 h of testing, the number of defects became physically uncountable. Size 4 and 2 blisters with dense or medium-dense intensities were also observed during this period. Both ASP and UM were removed from the test program after 4,320 h of testing.

UM (see figure 52) exhibited the worst surface failures compared to the rest of the coating systems. It had two defects initially, and this number rapidly increased after 1,440 h of laboratory testing. Numerous blisters were observed, and the panels had G4 to G1 rusting. The rust pits and blisters covered almost entire panel surface, and all of the blisters were filled with rust. The thin DFT of UM could be one of the reasons for the severe surface failures.

The two-coat (see figure 23) and EM (see figure 31) coating systems reacted differently compared to ASP and UM. Both systems did not develop any surface blisters or rust pits during the entire test period. However, when panels from the two-coat system were scanned by the holiday detector after 6,120 h, all of the test panels suggested the development of numerous defects by emitting a countless beeping sound. Most of the EM panels also indicated many defects after 6,480 h. Microscopic examination revealed numerous hairline cracks that had developed on the surface of the two-coat test panels. Figure 60 shows the photomicrograph of surface cracking of the two-coat panels. EM, however, did not show any surface cracking/ deterioration when examined under the optical microscope.

It was difficult to explain the cracking phenomena of the two-coat system using the information obtained from this study; however, one hypothesis could be made. Hare described the relationship of adhesion and cohesion with the internal stress in *Paint Film Degradation*.[35] According his theory:

> The adhesion and cohesion strengths maintain the integrity of the coating film. The internal stresses arising from solvent loss, polymerization of the binder, and from film formations are always counterproductive to good mechanical properties. When forces from internal stress are larger than film's cohesive strength, the film cracks on the surface. In most cases, the internal stress stored within the film minimizes the system's ability to accommodate additional tensile stress from external sources, or from the internal stress produced by long- term aging. (pp 142)

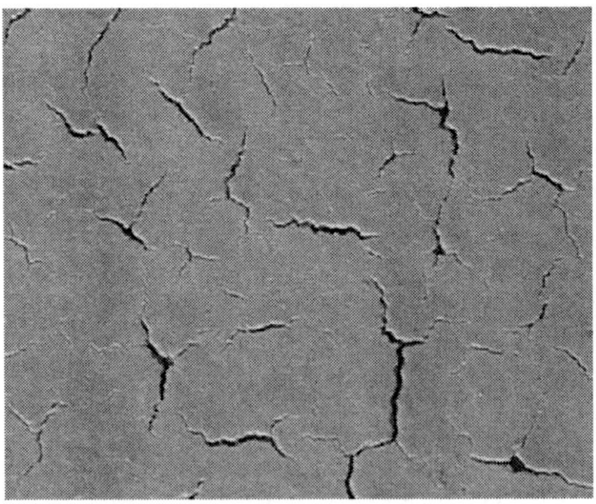

Figure 60. Photo. Surface coating failure by cracking (two-coat system).

Hare described that the external stress from the service includes bending, abrasion, impact, etc., as well as the hygrothermal gradients. He also indicated that over time, some polymers will undergo substantial polymerization and cross linking after film formation, particularly in the presence of UV light, and consequently increase the internal stress.[35] Some of the observed cracking in the present two-coat system could be explained based on this theory.

The microscopic examination of the three-coat system taken after 6,120 h also revealed some small holes on the panel surfaces. Figure 61 shows the photomicrographs of the surface of the three-coat system. However, the micro-sized holes were not holidays and thus did not affect the coating performance during ALT.

Outdoor Exposure Testing

Most coating systems performed well after 24 months in ME. The three-coat (see figure 20), two-coat (see figure 24), ASP (see figure 28), HBAC (see figure 41), and HRCSA (see figure 36) systems did not developed any surface failures and developed only a few defects, blisters, and rust pits on some test panels. EM did not exhibit any surface blisters and rust pits but developed some holidays after 24 months of exposure in ME. SLX (see figure 49) had one large rusted blister (larger than size 2) on one of the test panels and developed four defects during exposure in ME. Performance of the WBEP panels (see figure 45) was compromised by defects formed from coating application. Certain areas of the film might not have formed uniformly, causing areas of holidays on the panels. Some pinholes were also observed on the test panels, which developed into rusted blisters.

WBEP exhibited F6 blistering, G9 rusting, and had 13 holidays detected by the holiday detector. Although UM (see figure 53) developed F8 and M8 blistering, G8 to G5 rusting, and several holidays that were detected by the holiday detector, the surface failure after ME was much less severe compared to its surface failure in ALT.

All coating systems performed well in NW and NWS except for WBEP (see figure 46 and figure 47). As discussed earlier, WBEP test panels had some initial film defects. DFT was less than 1 mil (25.4 μm) at several areas. There were also some small dent areas on the surface. Typical defective surface condition and the resultant surface appearance after outdoor exposure can be seen in figure 62. The small dents grew into size 6 blisters after NW. In NWS, the small dents grew into size 6 rusted blisters, and all of the thin DFT areas became rusted (the rusting grade was G6). Many defects were detected by the holiday detector on the test panel surfaces.

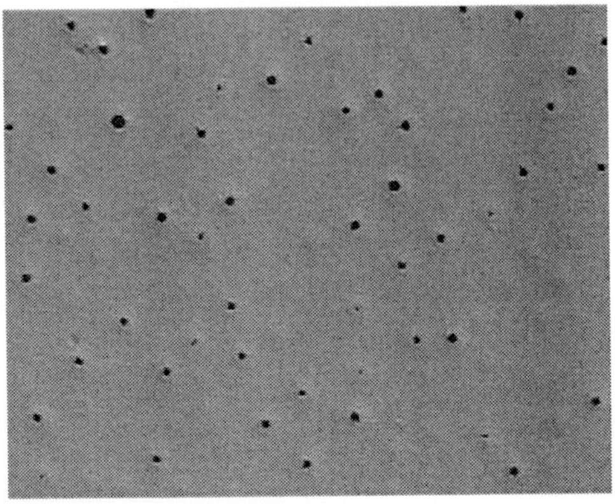

Figure 61. Photo. Surface condition of the three-coat system.

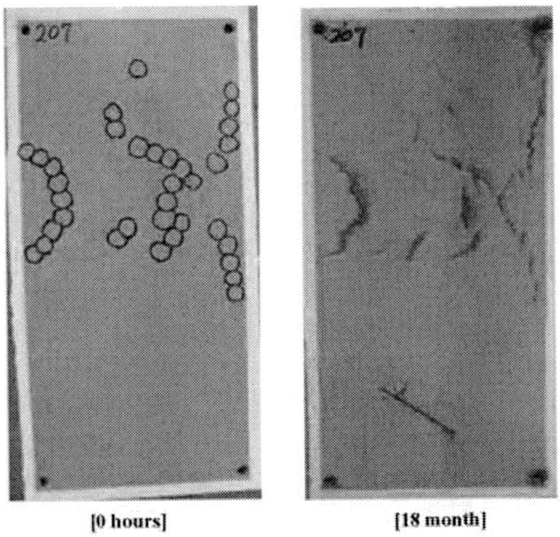

Figure 62. Photo. Large panels with defects from coating application deficiency in NWS.

To summarize, UM and WBEP had more surface failures than the other coating systems in outdoor exposures.

Growth of Rust Creepage

All of the coating systems developed some degree of rust creepage at the scribe line. Table 16 shows the creepage measurement of each coating system at the completion of each test method. Figure 63 through figure 66 show the mean creepage growth with time for each coating system in ALT, ME, NW, and NWS exposure conditions, respectively.

Table 16. Average rust creepage developed

Coating System	ALT		Rust Creepage (mm)		
	Time Exposure (hours)	Rust Creepage (mm)	ME (24 months)	NW (18 months)	NWS (18 months)
Three-coat	6,840	5.3	0	0	0.5
Two-coat	6,840	4.7	1.6	1.6	1.5
ASP	4,320	6.8	1.8	0	0
EM	6,840	6.5	0.9	0.6	1.6
HRCSA	6,840	0.7	1	0.7	0.7
GFP	6,840	7.1			
HBAC	5,040	9.3	1.3	0	3.7
WBEP	5,040	15.9	1.1	0.6	2.3
SLX	4,320	21.9	30.5	2.2	12.5
UM	4,320	35.6	5.2	0.7	6.6

1 inch = 25.4 mm.

Note: The blank cells indicate that no outdoor exposure data were available.

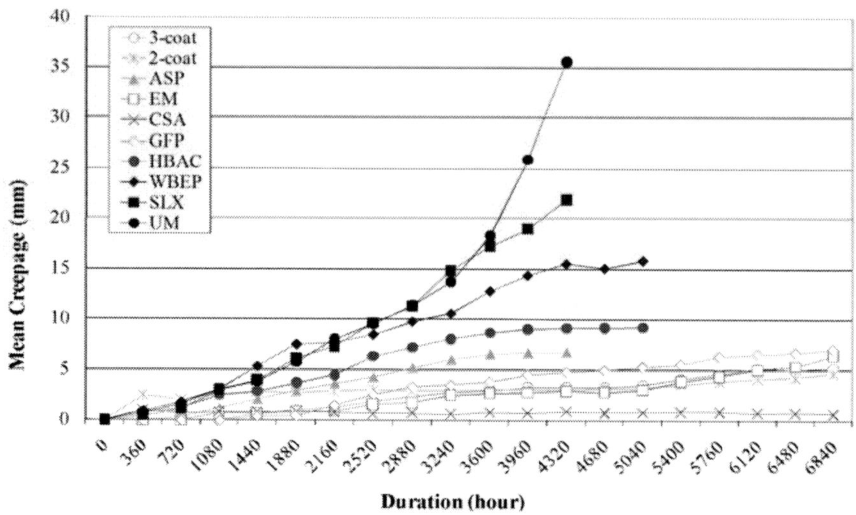

1 inch = 25.4 mm.

Figure 63. Graph. Development of rust creepage during ALT.

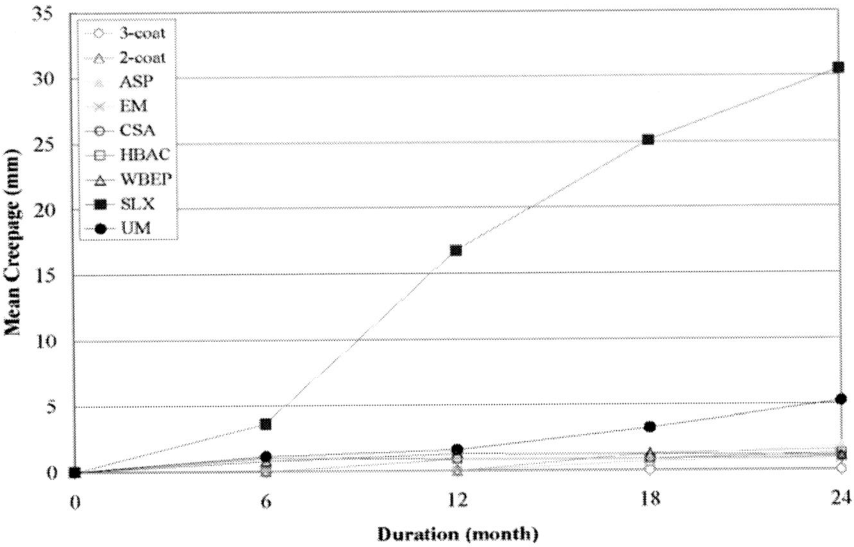

1 inch = 25.4 mm.

Figure 64. Graph. Development of rust creepage during ME.

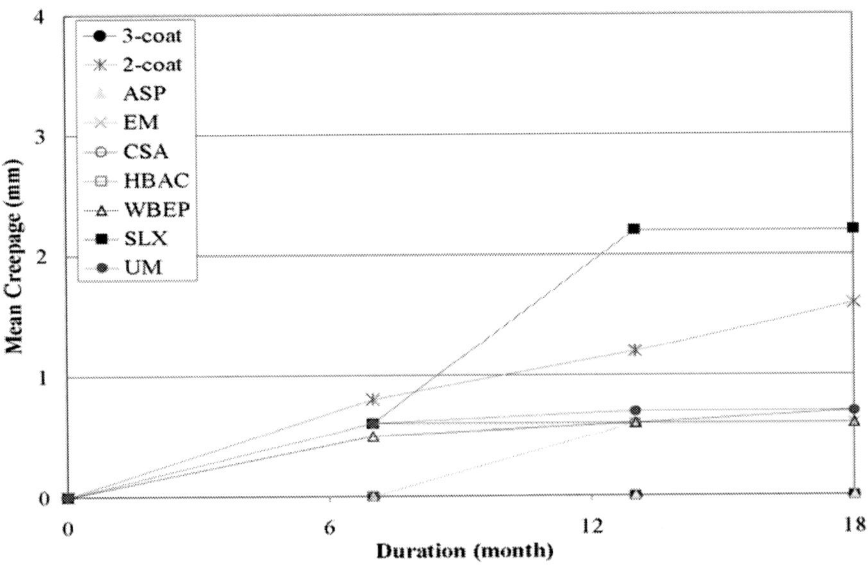

1 inch = 25.4 mm.

Figure 65. Graph. Development of rust creepage during NW.

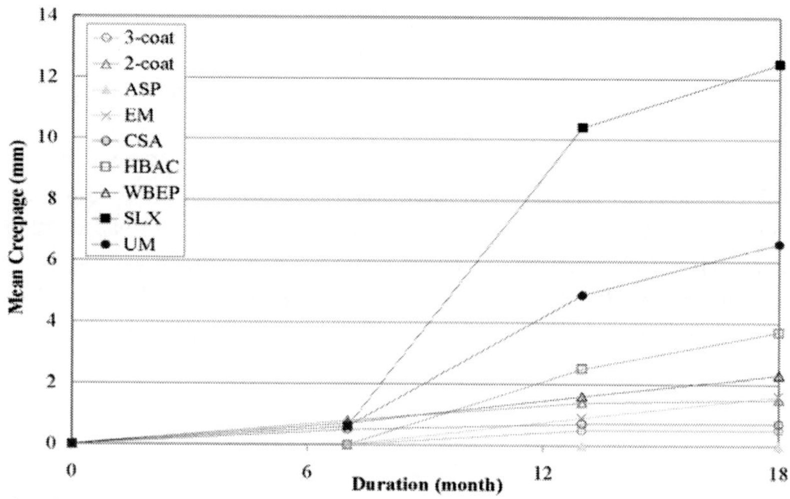

1 inch = 25.4 mm.

Figure 66. Graph. Development of rust creepage during NWS.

ALT

After ALT, the three-coat and two-coat control systems had moderate creepage at 0.21 and 0.18 inches (5.3 and 4.7 mm), respectively. The three-coat system started to show creepage after 1,080 h of incubation time, while the two-coat system developed rust creepage along the line after 360 h (one test cycle) and slowly grew to 0.18 inches (4.7 mm) after 6,840 h.

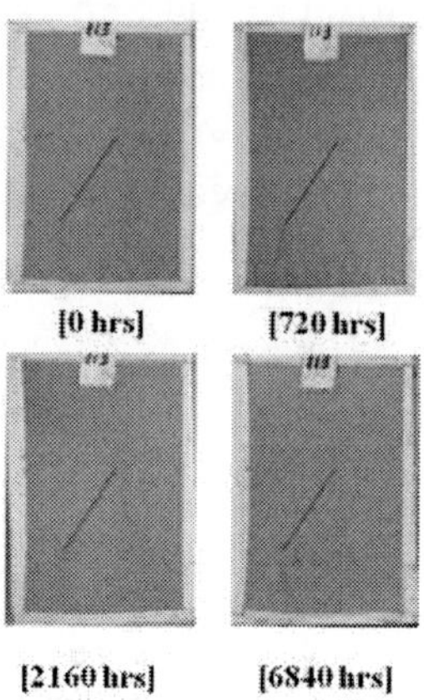

Figure 67. Photo. Rust creepage of HRCSA during ALT.

HRCSA had the lowest rust creepage of 0.03 inches (0.7 mm) at the completion of ALT. Figure 67 shows the creepage growth during ALT. HRCSA developed initial creepage after 360 h; however, the creepage did not grow much during the entire test period.

EM and GFP had similar performance, with final rust creepage of 0.25 and 0.28 inches (6.5 and 7.1 mm) after incubation times of 1,080 and 1,440 h, respectively. ASP started developing creepage after 720 h and had a creepage of 0.27 inches (6.8 mm) when it was removed from the test program after 4,320 h due to the development of severe surface defects. HBAC and WBEP had creepage of 0.36 and 0.59 inches (9.3 and 15.1 mm), respectively, when they were removed from the test program after 5,040 h. Both coatings started developing rust creepage after the first test cycle of 360 h.

UM had the highest creepage of 1.37 inches (35 mm), followed by SLX, which had a final creepage of 0.98 inches (25 mm) after the first test cycle of 360 h. They were removed from the test program after 4,320 h.

In summary, HRCSA outperformed the three-coat and two-coat control systems in terms of rust creepage. EM and GFP had more creepage than the two controls. All other one-coat systems had much larger rust creepage than the two controls.

Rust creepage is typically produced as a result of steel dissolution in corrosive environments at the scribed area.[36] The primers in the coating systems prevented the steel corrosion through cathodic protection or inhibitive pigments. Zinc-rich primers of the two controls provided sacrificial cathodic protection for the steel panel. It was interesting to note from the major element contents of the pigments (see table 5) that five of the eight one-coat systems which had an elevated amount of zinc and some amount of phosphorus did not perform as well as the two one-coat systems without any zinc and phosphate. Specifically, GFP and EM outperformed the other one-coat systems except for HRCSA.

Outdoor Exposure Testing

Most coating systems developed minimal rust creepage compared to those in ALT after 24 months of exposure in ME and 18 months exposure in NW and NWS.

The three-coat system did not develop rust creepage at the scribe at the end of ME as well as exposure in NW for 18 months. However, 0.02 inches (0.5 mm) of creepage was observed after 13 months in NWS. FHWA's prior coating test results showed that the three-coat system with a zinc-rich primer usually did not develop creepage at the scribe when exposed to a salt-rich ME.

The 0.02-inch (0.5-mm) rust creepage development indicated that daily spray of 15 percent salt solution is a more severe exposure condition than ME.

The two-coat system exhibited less than 0.08 inches (2 mm) of creepage in all three outdoor exposures. It showed visible creepage at the scribe after 18 months in ME and after 6 months in NW and NWS.

ASP did not develop rust creepage in NW and NWS but started to develop creepage of 0.07 inches (1.8 mm) in ME during the first 6 months.

EM started to show rust creepage after 18 months of exposure in ME and after 13 months of exposure in NW and NWS. The system had less than 0.08 inches (2 mm) of mean final creepage in all three outdoor tests.

HRCSA and WBEP exhibited visible rust creepage after 6 months exposure in three outdoor conditions. However, rust creepage did not grow significantly during the test period. HRCSA had 0.039 inches (1 mm) of creepage in three outdoor exposures at the end of the test period. WBEP demonstrated the highest creepage of 0.09 inches (2.3 mm) in NWS, 0.04 inches (1.1 mm) in ME, and 0.02 inches (0.6 mm) in NW.

HBAC did not show creepage in NW and had 0.039 inches (1 mm) of creepage in ME. However, it had around 0.16 inches (4 mm) of creepage in NWS.

SLX performed poorly in ME and NWS. The final mean creepage was 1.12 and 0.51 inches (31 and 13 mm), respectively. SLX performed very well in the NW with creepage of 0.08 inches (2 mm).

At the completion of the test, UM had creepage values of 0.20 and 0.27 inches (5 and 7 mm) in ME and NWS, respectively. The creepage developed in NW was less than 0.039 inches (1 mm).

3.3. Correlation among Performance Parameters and Exposure Conditions

Correlation among Characterization Parameters in a Specific Exposure Condition

Linear Regression Analysis

Linear regression analysis was performed to identify relationships between the various performance characterization parameters and also to establish if correlations exist between the various exposure conditions involved. All parameters and the corresponding numerical values of one-coat systems and the two controls toward the end of the test period for each exposure environment are shown in table 17.

Table 17. Summary of ALT and outdoor performance data

Coating System	**Three-Coat**	**Two-Coat**	**ASP**	**EM**	**HRCSA**	**GFP**	**HBAC**	**WBEP**	**SLX**	**UM**
ALT										
Gloss reduction percent	50.9	60.3	27.6	99	66.7	41.6	79.5	77.8	18.5	23.8
Color reduction (ΔE)	1.2	1.4	1.4	8.6	6.3	8.2	10.9	4.7	3.1	3.7
Variation in adhesion strength: scribed (psi)	-12	107	4	-23	13	30	-12	-32	-22	-22
Variation in adhesion strength: unscribed (psi)	-37	60		-34	11	20	75	-46	-44	
Number of coating defects	1	550	200	100	2	0	7	5	32	550
Rust creepage at the scribe (mm)	5.3	4.7	6.8	6.5	0.7	7.1	9.3	15.9	21.9	35.6
ME										
Gloss reduction percent	28.9	91.5	52.7	97.7	30.6		29.2	66.9	32.8	4.3
Color reduction (ΔE)	1	3.5	1.6	9.6	9.8		2.2	1.9	0.4	0.4
Variation in adhesion strength: scribed (psi)	-3	10	-8	-17	17		8	-2	-18	-10
Variation in adhesion strength: unscribed (psi)	-4	10	-6	-2	9		53	1	-50	-19
Number of coating defects	1	0	0	35	0		0	13	11	550
Rust creepage at the scribe (mm)	0	1.6	1.8	0.9	1		1.3	1.1	30.5	5.2
NW										
Gloss reduction percent	29.5	39	10.1	96.9	81.9		24.4	59.3	20.6	1.5
Color reduction (ΔE)	1	0.5	0.3	14.4	6.3		3.3	1.5	0.8	0.2

Coating System	Three-Coat	Two-Coat	ASP	EM	HRCSA	GFP	HBAC	WBEP	SLX	UM
Variation in adhesion strength: scribed (psi)	3	46	34	3	20		85	-21	4	-5
Variation in adhesion strength: unscribed (psi)	-10	35	39	5	16		25	-3	-17	-11
Number of coating defects	1	1	4	12	1		1	39	68	67
Rust creepage at the scribe (mm)	0	1.6	0	0.6	0.7		0	0.6	2.2	0.7
NWS										
Gloss reduction percent	14.2	34.5	15	97.3	74.1		16.5	63.8	12.4	0.5
Color reduction (ΔE)	1	0.3	0.4	15.3	8.2		3.3	1.7	0.4	
Variation in adhesion strength: scribed (psi)	18	23	20	-2	29		36	-18	-15	-6
Variation in adhesion strength: unscribed (psi)	5	31	23	-3	24		87	8	3	7
Number of coating defects	0	0	0	12	0		0	174	28	30
Rust creepage at the scribe (mm)	0.5	1.5	0	1.6	0.7		3.7	2.3	12.5	6.6

1 inch = 25.4 mm.
1 psi = 6.89 kPa.
Note: The blank cells indicate that no outdoor exposure data were available.

Combinations of variable pairs for color, gloss, adhesion strength of scribed and unscribed panels, number of coating defects, and rust creepage at the scribe for each individual exposure condition were used for a linear regression analysis. Exposure conditions were ALT, ME, NW, and NWS. The combinations of the various parameters are as follows:

- Color versus gloss.
- Color versus rust creepage.
- Color versus coating defects.
- Gloss versus coating defects.

- Gloss versus rust creepage.
- Coating defects versus rust creepage.
- Adhesion strength (scribed) versus rust creepage.
- Adhesion strength (unscribed) versus rust creepage.
- Adhesion strength (scribed) versus coating defects.
- Adhesion strength (unscribed) versus coating defects.

Linear regression analysis of the above combinations of variables was conducted using Microsoft® Excel, and the corresponding R-squared values were recorded (see table 18). Correlations with R-squared values higher than 0.6 were identified and further explored for a better numerical relationship. Examples of a good and poor correlation using linear regression analysis are shown in figure 68 and figure 69, respectively. Regression analysis of color versus gloss in NW and NWS resulted in R-squared values greater than 0.69. Another promising correlation was found between adhesion strength of unscribed panels and coating defects, which had an R-squared value of 0.51 in NW. Figure 69 shows the correlation between color and gloss in ME with an R-squared value of 0.193, indicating a poor correlation.

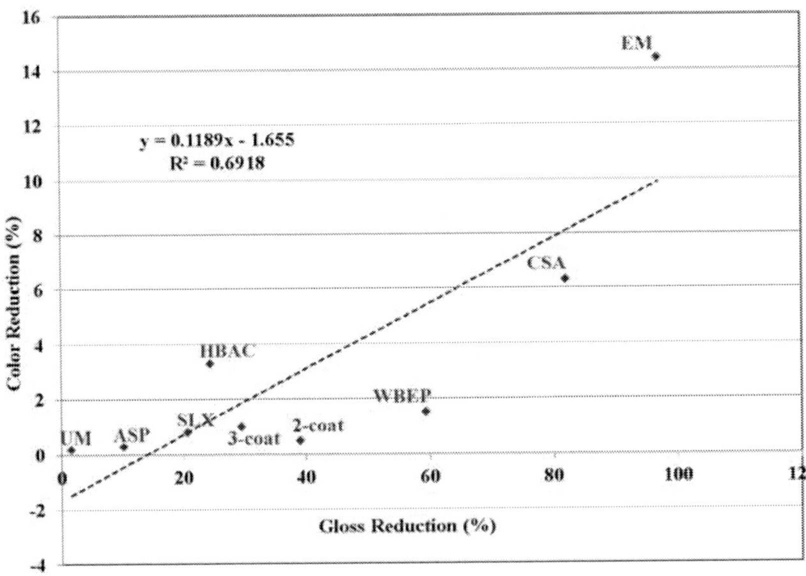

Figure 68. Graph. Positive linear regression analysis between color and gloss in NW.

Table 18. R-squared values from linear regression analysis for various performance parameter combinations

Combination of Parameters	R-squared Values				
	ALT	ME	NW	NWS	Overall Correlation
Color versus gloss	0.328	0.193	0.692	0.711	0.015
Color versus rust creepage	0.021	0.121	0.015	0.053	0.101
Color versus coating defects	0.191	0.075	0.077	0.021	0.063
Gloss versus coating defects	0.084	0.242	0.124	0.006	0.063
Gloss versus rust creepage	0.248	0.060	0.000	0.166	0.088
Coating defects versus rust creepage	0.167	0.001	0.278	0.009	0.088
Adhesion strength (scribed) versusrust creepage	0.193	0.255	0.038	0.305	0.088
Adhesion strength (unscribed) versusrust creepage	0.199	0.105	0.068	0.018	0.087
Adhesion strength (scribed) versus coating defects	0.233	0.073	0.324	0.453	0.063
Adhesion strength (unscribed) versus coating defects	0.190	0.070	0.511	0.074	0.063

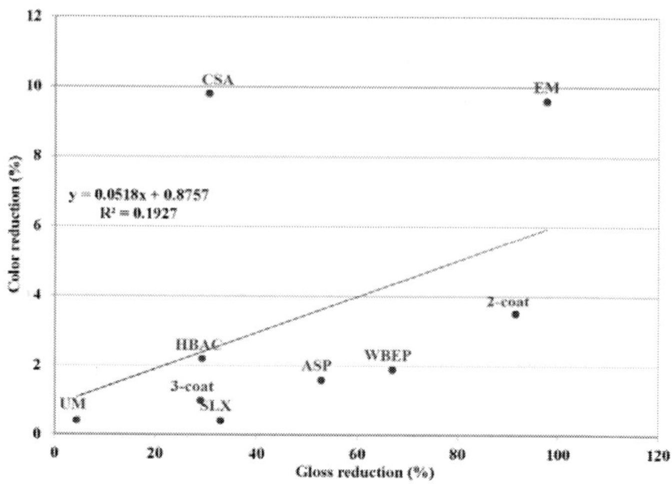

Figure 69. Graph. Poor linear regression analysis between color and gloss in ME.

As seen in figure 68, the relationship between color and gloss reduction is not a realistic model because the line intercepts the negative y-axis between zero and -2. This means that for zero gloss reduction, the reduction in color is a negative value. However, the relationship indicates that a correlation exists between color and gloss reduction for one-coat systems in NW. A better correlation can be established through regression analysis using an exponential, power, or polynomial fit instead of a linear fit.

Regression Analysis Using Exponential and Polynomial Fits

After the above correlations yielded encouraging results from regression analysis using a linear fit, the relationships of color versus gloss and adhesion strength versus coating defects were correlated using an exponential fit.

The corresponding exponential fit resulted in an increased R-squared value of 0.70 for color versus gloss. For simplicity, outdoor exposure data in both NW and NWS were pooled together for the regression analysis. The larger dataset pooled from NW and NWS has increased sample size, resulting in a statistically improved correlation with a lower standard deviation. The resulting correlation is shown in figure 70. True performance of various one-coat systems can be gauged by excluding the control systems. The correlation between color and gloss without the controls resulted in an increased R-squared value from 0.70 to 0.77 as shown in figure 71.

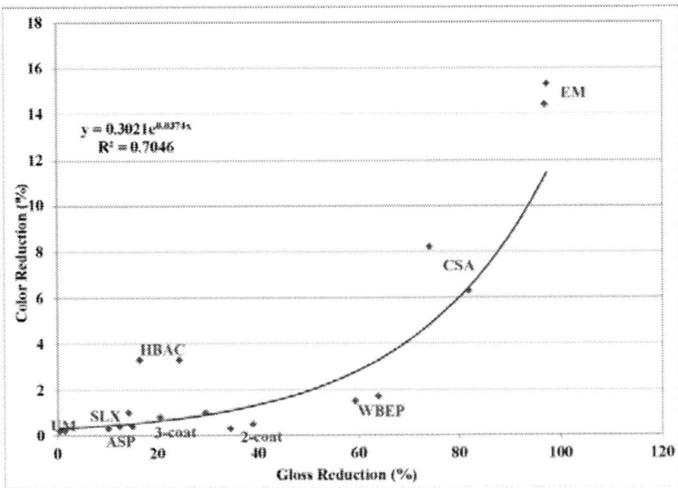

Figure 70. Graph. Regression analysis of color versus gloss for one-coat and control coating systems in NW and NWS.

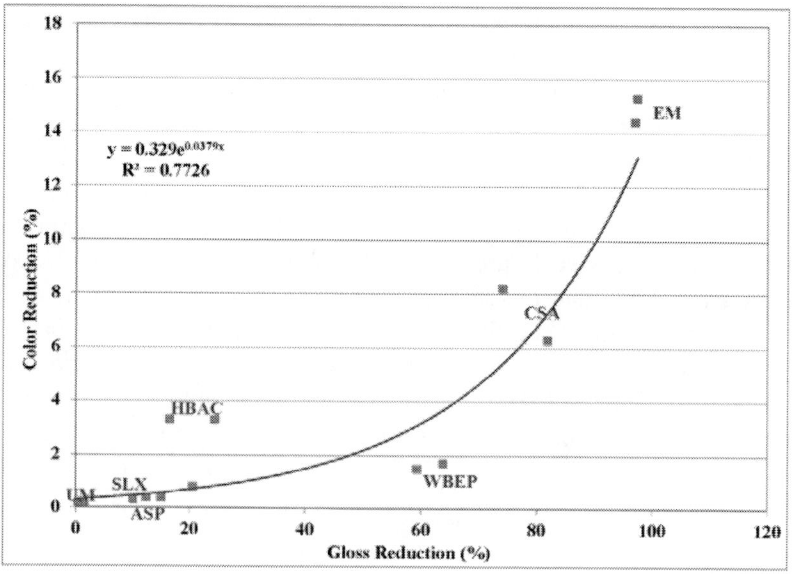

Figure 71. Graph. Improved regression analysis results from figure 70.

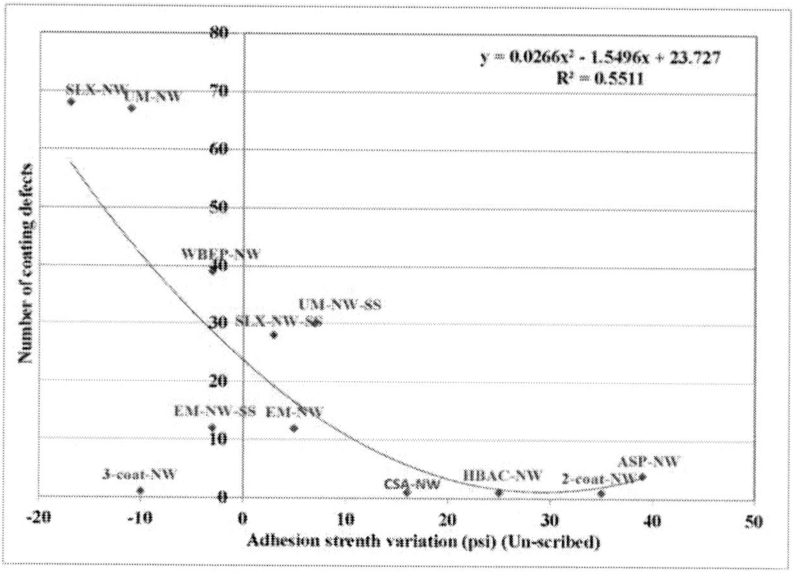

Figure 72. Graph. Regression analysis of adhesion strength versus coating defects for one- coat and control coating systems in NW.

Linear regression analysis of adhesion strength variation in unscribed panels correlated with coating defects, yielding an R-squared value of 0.51. The adhesion and coating defects data of all one-coat systems along with the control coating systems for both NW and NWS were pooled together, and a regression analysis was performed with a polynomial fit to yield an increased R-squared value of 0.56.

When the control coating systems were not included, the R-squared value increased from 0.56 (see figure 72) to 0.82 (see figure 73). This improved correlation between adhesion strength variations and the number of coating defects as well as color and gloss indicates that these relationships are more likely characteristics of one-coat systems. Coating systems with zero defects and numerous holidays (>100 and physically impossible to count) on the surface were not included in this analysis. These included ASP, HRCSA, and HBAC.

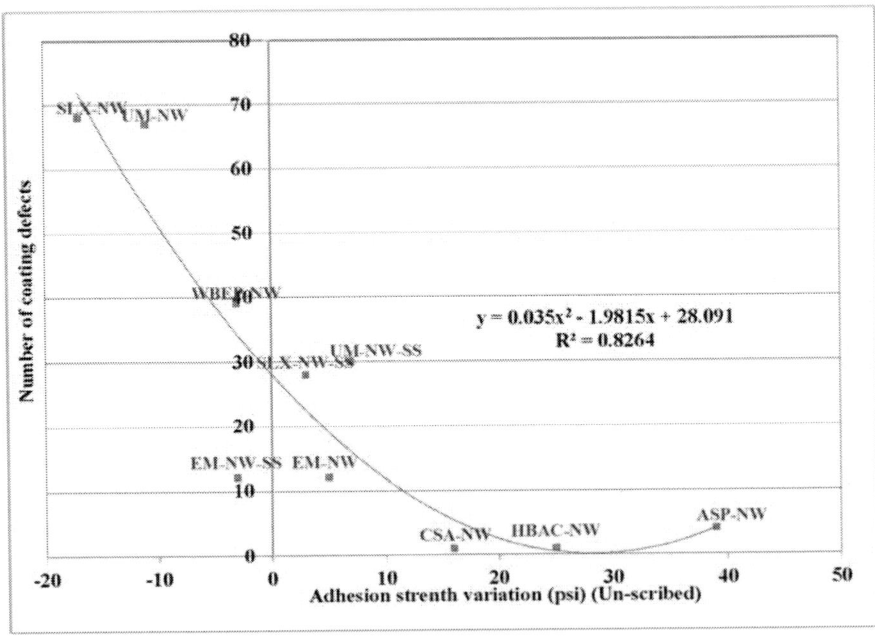

Figure 73. Graph. Improved regression analysis result of adhesion strength versus coating defects for one-coat systems in NW.

Correlation among Exposure Conditions for a Specific Characterization Parameter

Linear Regression Analysis

Linear regression analysis was performed for variations in individual performance parameters, such as color and gloss, to examine the relationship among ALT, ME, NW, and NWS.

A combination matrix showing the performance parameters and the various exposure conditions involved is shown in table 19. Regression analysis of these exposure condition combinations for the performance evaluation variables resulted in R-squared values shown in table 20.

Table 19. Linear regression analysis combinations of exposure conditions

Characterization Parameter	Combination 1	Combination 2	Combination 3	Combination 4	Combination 5	Combination 6
Color	ALT versus ME	ALT versus NW	ALT versus NWS	ME versus NW	ME versus NWS	NW versus NWS
Gloss	ALT versus ME	ALT versus NW	ALT versus NWS	ME versus NW	ME versus NWS	NW versus NWS
Adhesion strength scribed	ALT versus ME	ALT versus NW	ALT versus NWS	ME versus NW	ME versus NWS	NW versus NWS
Adhesion strength unscribed	ALT versus ME	ALT versus NW	ALT versus NWS	ME versus NW	ME versus NWS	NW versus NWS
Coating defects	ALT versus ME	ALT versus NW	ALT versus NWS	ME versus NW	ME versus NWS	NW versus NWS
Rust creepage	ALT versus ME	ALT versus NW	ALT versus NWS	ME versus NW	ME versus NWS	NW versus NWS

Table 20. R-squared values of linear regression analysis of exposure conditions

Combination of Parameters	ALT versus ME	ALT versus NW	ALT versus NWS	ME versus NW	ME versus NWS	NW versus NWS
Color	0.231	0.417	0.407	0.731	0.792	0.988
Gloss	0.330	0.635	0.613	0.337	0.411	0.961
Adhesion strength scribed	0.302	0.199	0.243	0.262	0.531	0.668
Adhesion strength unscribed	0.576	0.834	0.767	0.367	0.672	0.351
Coating defects	0.391	0.004	0.037	0.371	0.001	0.185
Rust creepage at the scribe	0.214	0.078	0.514	0.569	0.384	0.463

Regression Analysis Using Power and Polynomial Fit

Linear regression analysis for a linear fit yielded favorable R-squared values for all performance parameters in various exposure condition combinations except for the number of coating defects developed and rust creepage at the scribe. R-squared values higher than 0.65 were chosen and based on favorable correlations from the linear fit, and a regression analysis was performed using a power fit.

Figure 74 shows the relationship between gloss changes in NW and NWS that yielded an R-squared value of 0.94. Similarly, figure 75 shows the relationship between color variations in NW and NWS regressed using the power fit, resulting in an R-squared value of 0.96. Regression analysis of variation in adhesion strength for scribed panels was performed using a polynomial equation of 2d order with an R-squared value of 0.75. Figure 76 shows the analysis results.

Figure 77 shows the strong relationships among adhesion strength variations of scribed panels for the following:

- ALT (dependent) versus NW (independent).
- ALT (dependent) versus NWS (independent).
- ME (dependent) versus NWS (independent).

ASP and UM did not have any changes in adhesion strength toward the end of the testing period. As a result, the values of these adhesion strength variations were not used in the regression analysis.

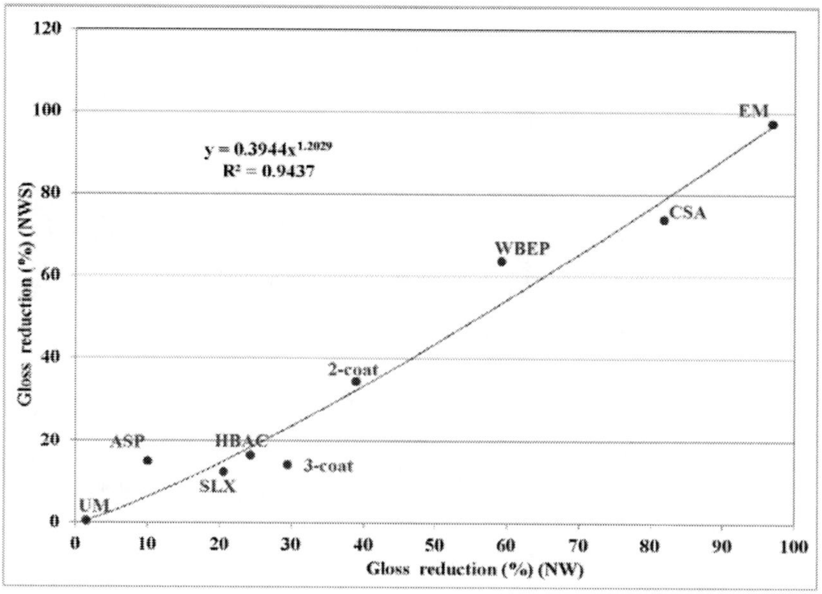

Figure 74. Graph. Gloss reductions in NW versus gloss variations in NWS.

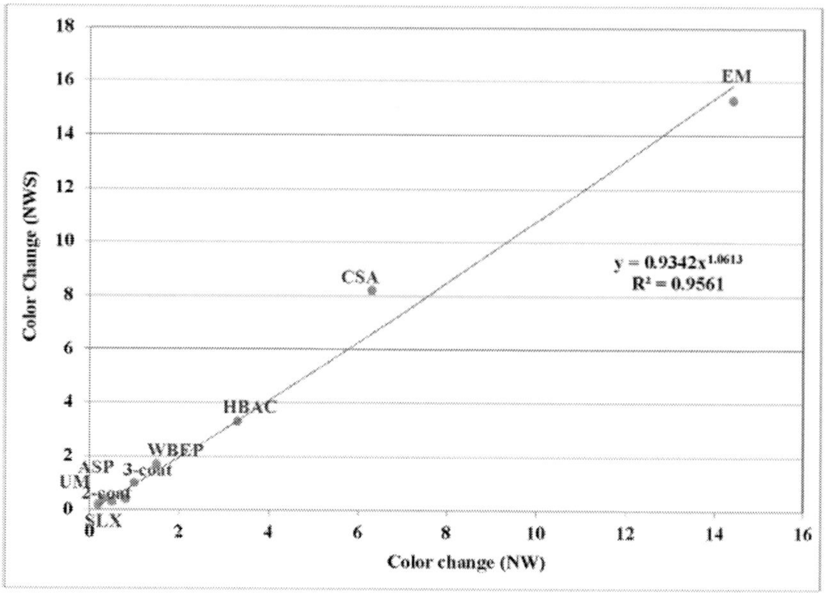

Figure 75. Graph. Color variations in NW versus NWS.

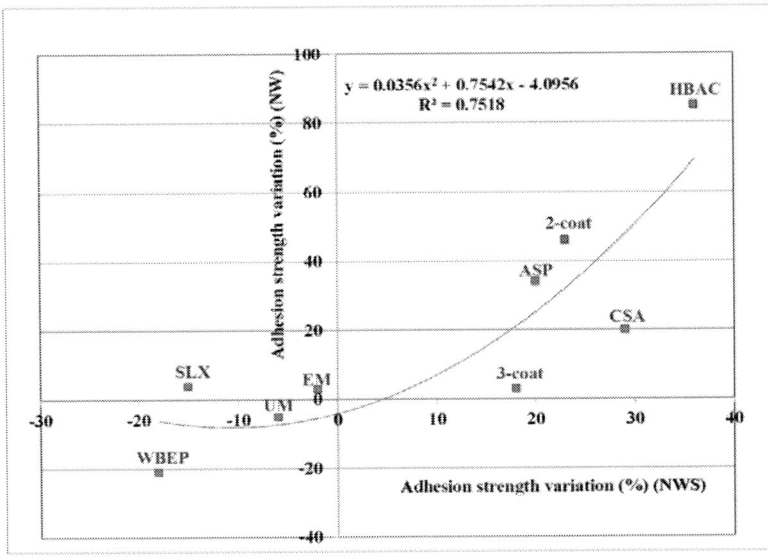

Figure 76. Graph. Relationship between adhesion strength variations of scribed panels in NW and NWS.

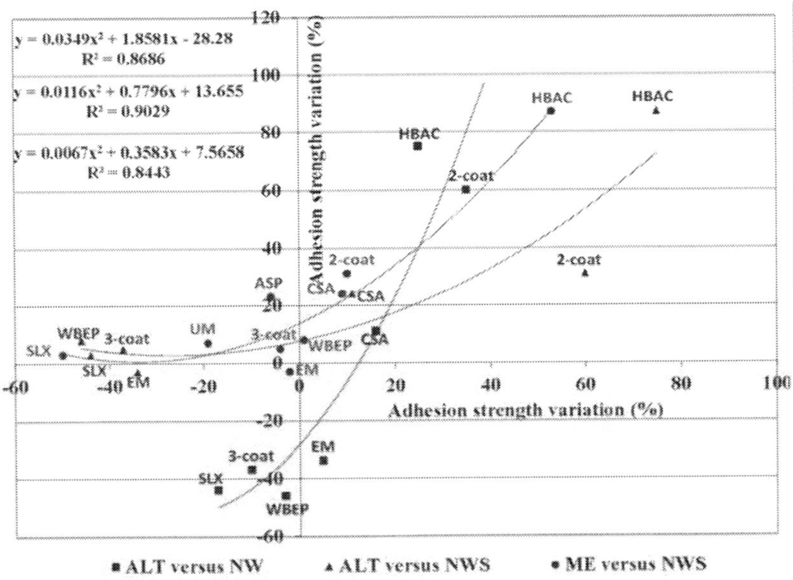

Figure 77. Graph. Relationship between adhesion strength variations of unscribed panels.

Summary of Relationship between Variables and Exposure Conditions

Table 20 summarizes the final R-squared values for favorable correlations among the performance parameters and exposure conditions.

3.4. Comprehensive Performance Evaluation

Gloss and Color

All coating systems had similar gloss reduction in NW and NWS. In general, the salt solution spray had no effect on gloss reduction. ME had a larger impact on gloss reduction compared to NW and NWS.

EM had higher than 95 percent gloss reduction in both ALT and the outdoor exposures. The two- coat system, WBEP, and ASP had the highest gloss reduction in ME, while HBAC, HRCSA, and WBEP had the highest gloss reduction in ALT. HRCSA had the highest gloss reduction in NW and NWS.

The color changes of the coating systems were dissimilar in both ALT and the outdoor exposures. For example, WBEP and HBAC had large ΔE values in ALT compared to ME, NW, and NWS.

UM and SLX exhibited strong UV resistance in both laboratory and outdoor exposures, as demonstrated by the low color and gloss reduction values. The two-coat system and ASP showed poor UV resistance in ME. EM and HRCSA exhibited poor UV resistance in all environments.

Presence of AR compounds can result in reduced weathering performance in outdoor exposure conditions since UV light causes modified surface appearance of AR coatings due to yellowing and/or chalking.

The binder of several coating systems (i.e., three-coat, EM, GFP, HBAC, WBEP, SLX, and UM) consisted of some degrees of aromaticity, which typically reduces the weatherability of these coating systems. The AR/AP ratio of all one-coat systems in table 6, when correlated with gloss reduction, demonstrated that higher AR/AP results in higher gloss reduction (figure 78).

Pencil Scratch Hardness

The coating hardness changes of the 10 coating systems after outdoor exposure were very similar to that after laboratory testing. ASP, SLX, and UM became harder after ALT and the outdoor exposures. The two-coat system, which did not exhibit a hardness change in ALT, had an increased hardness by about 1 degree in the outdoor exposure tests. The remaining six coating systems retained the same pencil scratch hardness.

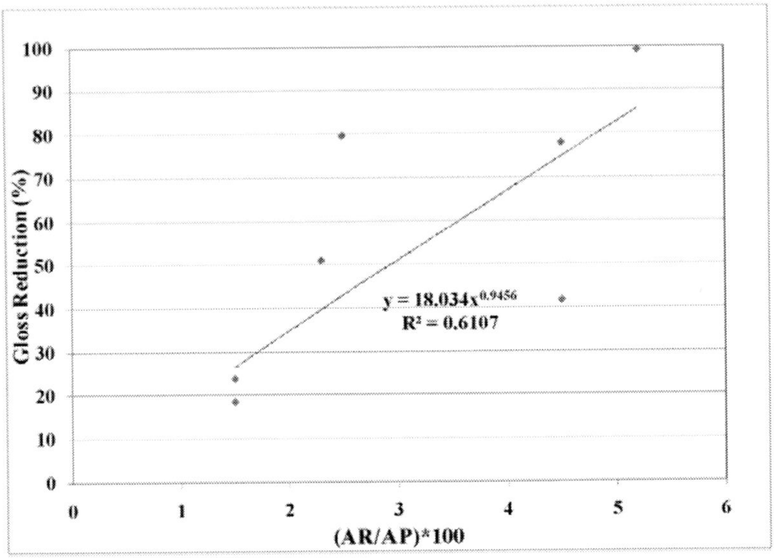

Figure 78. Graph. Gloss reduction as a function of aromaticity.

Adhesion Strength

There were no significant adhesion strength changes for most coating systems except for SLX (ME), HBAC (all exposure conditions), and the two-coat system (ALT). A lack of significant decrease in adhesion strength after testing for any particular coating was observed.

Surface Appearance and Failure

As expected, coating systems developed more surface failures in ALT than in the outdoor exposures. UM had severe surface failures after ALT, moderate surface failure after exposure to ME, and minimal surface failure in both NW and NWS.

ASP had severe surface failure in ALT but no surface failure in all three outdoor exposures. The two-coat system developed surface film cracking during ALT but not in other outdoor exposure conditions. EM had some invisible defects detected by a holiday detector after ALT, fewer defects after ME, and no defects in NW and NWS.

Rust Creepage

For most coating systems, the creepage developed in outdoor exposure was much smaller than the creepage developed in ALT for all of the coating systems. HRCSA had less than 0.039 inches (1 mm) of creepage in all tests.

SLX had about 1.21 inches (31 mm) of creepage in ME exposure, which was 0.35 inches (9 mm) greater than the rust creepage developed in ALT. This confirmed that three outdoor exposure conditions were milder than ALT for all coating systems except SLX.

Performance Ranking

All one-coat systems were ranked based on their performance in ALT and the three outdoor exposures. The total time of exposure conditions are as follows:

- ALT: 6,840 h.
- ME: 17,520 h.
- NW: 13,140 h.
- NWS: 13,140 h.

Each ALT cycle consisted of 360 h, while the outdoor exposures were continuous exposures for 24 months (ME) and 18 months (NW and NWS). Performance data can be weighed equally by calculating the rate of change of the variables per each cycle. The number of equivalent cycles for all exposure conditions based on a 360-hr cycle are as follows:

- ALT: 19 cycles.
- ME: 49 cycles.
- NW: 37 cycles.
- NWS: 37 cycles.

Weight of Exposure Conditions

It is important to weigh the exposure conditions based on their impact on the performance of the coating systems. Rust creepage, the number of coating defects, rusting and blister grades, color variation, gloss reduction, and variation in adhesion strength are quantitative parameters that indicate the impact of an exposure condition. Quantitative analysis was used to calculate the coefficient of impact of each exposure condition.

Rust creepage is a dynamically changing parameter, and the rate of change of this variable can be calculated from creepage data. *Rate of creepage* can be defined as the change in the value of creepage with time.

$$\text{Rate of creepage} = \frac{\Delta\ \text{creepage}}{\Delta\ \text{time}} \tag{2}$$

where Δ indicates a change in the parameter.

The rate of creepage can be obtained from the slope of the linear fit of creepage plotted as a function of time. All parameters other than creepage are snapshots at specific periods of time of exposure in all test conditions, and their rate of variation can be calculated per cycle. For example, the number of coating defects developed per cycle, the percentage of gloss reduction per cycle, color variation per cycle, and adhesion strength variation per cycle can be calculated for each coating system in all exposure conditions using the total time of exposure.

Linear regression analysis was used to fit creepage as a function of time, and the corresponding slope was obtained as the rate of creepage. Normalized values of rust creepage (rate), coating defects per cycle, color variation per cycle, gloss reduction per cycle, and adhesion strength variation per cycle were calculated for all coating systems in ALT, MW, NW, and NWS, and average values of each normalized parameter were calculated. Average coating defects, rust creepage, color, and gloss reductions are shown in table 21 through table 24.

Table 21. Average coating defects developed

Coating System	ALT		ME		NW		NWS	
	Number of Coating Defects	Coat Defects/ Cycle	Number of Coating Defects	Coat Defects/ Cycle	Number of Coating Defects	Coat Defects/ Cycle	Number of Coating Defects	Coat Defects/ Cycle
Three-coat	1	0.05	1	0.02	1	0.03	0	0.00
Two-coat	100	5.26	0	0.00	1	0.03	0	0.00
ASP	100	8.33	0	0.00	4	0.11	0	0.00
EM	100	5.26	35	0.72	12	0.33	12	0.33
HRCSA	2	0.11	0	0.00	1	0.03	0	0.00
HBAC	7	0.50	0	0.00	1	0.03	0	0.00
WBEP	5	0.36	13	0.27	39	1.07	100	2.74
SLX	32	2.67	11	0.23	68	1.86	28	0.77
UM	100	8.33	100	2.05	67	1.84	30	0.82
Average		3.43		0.37		0.59		0.52

Note: The blank cells indicate average parameter values used were based per cycle.

Table 22. Average rust creepage developed

Coating System	ALT		ME		NW		NWS	
	Rust Creepage at the Scribe (mm)	Creepage/Cycle	Rust Creepage at the Scribe (mm)	Creepage/ Cycle	Rust Creepage at the Scribe (mm)	Creepage/ Cycle	Rust Creepage at the Scribe (mm)	Creepage/ Cycle
Three-coat	5.3	0.28	0	0.00	0	0.00	0.5	0.01
Two-coat	4.7	0.25	1.6	0.03	1.6	0.04	1.5	0.04
ASP	6.8	0.57	1.8	0.04	0	0.00	0	0.00
EM	6.5	0.34	0.9	0.02	0.6	0.02	1.6	0.04
HRCSA	0.7	0.04	1	0.02	0.7	0.02	0.7	0.02
HBAC	9.3	0.66	1.3	0.03	0	0.00	3.7	0.10
WBEP	15.9	1.14	1.1	0.02	0.6	0.02	2.3	0.06
SLX	21.9	1.883	30.5	0.63	2.2	0.06	12.5	0.34
UM	35.6	2.97	5.2	0.11	0.7	0.02	6.6	0.18
Average		0.90		0.10		0.02		0.09

1 inch = 25.4 mm.

Note: The blank cells indicate average parameter values used were based per cycle.

Table 23. Average color reduction

Coating System	ALT		ME		NW		NWS	
	Color Reduction (CR) (ΔE)	CR/Cycle	CR (ΔE)	CR/Cycle	CR (ΔE)	CR/Cycle	CR (ΔE)	CR/Cycle
Three-coat	1.2	0.06	1	0.02	1	0.03	1	0.03
Two-coat	1.4	0.07	3.5	0.07	0.5	0.01	0.3	0.01
ASP	1.4	0.12	1.6	0.03	0.3	0.01	0.4	0.01
EM	8.6	0.45	9.6	0.20	14.4	0.39	15.3	0.42
HRCSA	6.3	0.33	9.8	0.20	6.3	0.17	8.2	0.22
HBAC	10.9	0.78	2.2	0.05	3.3	0.09	3.3	0.09
WBEP	4.7	0.34	1.9	0.04	1.5	0.04	1.7	0.05
SLX	3.1	0.26	0.4	0.01	0.8	0.02	0.4	0.01
UM	3.7	0.31	0.4	0.01	0.2	0.01	0.2	0.01
Average		0.30		0.07		0.09		0.09

Note: The blank cells indicate average parameter values used were based per cycle.

Table 24. Average gloss reduction

Coating System	Gloss Reduction (GR)Percent	GR/Cycle	GR Percent	GR/Cycle	GR Percent	GR/Cycle	GR Percent	GR/Cycle
Three-coat	50.9	2.68	28.9	0.59	29.5	0.81	14.2	0.39
Two-coat	60.3	3.17	91.5	1.88	39	1.07	34.5	0.95
ASP	27.6	2.30	52.7	1.08	10.1	0.28	15	0.41
EM	99	5.21	97.7	2.01	96.9	2.65	97.3	2.67
HRCSA	66.7	3.51	30.6	0.63	81.9	2.24	74.1	2.03
HBAC	79.5	5.68	29.2	0.60	24.4	0.67	16.5	0.45
WBEP	77.8	5.56	66.9	1.37	59.3	1.62	63.8	1.75
SLX	18.5	1.54	32.8	0.67	20.6	0.56	12.4	0.34
UM	23.8	1.98	4.3	0.09	1.5	0.04	0.5	0.01
Average		3.51		0.99		1.11		1.00

Note: The blank cells indicate average parameter values used were based per cycle.

An example for calculating the weighted average rate of the development of coating defects against a standard unit of 1 resulted in the following coefficients:

- ALT: 3.43/(3.43+0.37+0.59+0.52) = 0.70
- ME: 0.37/(3.43+0.37+0.59+0.52) = 0.07
- NW: 0.59/(3.43+0.37+0.59+0.52) = 0.12
- NWS: 0.52/(3.43+0.37+0.59+0.52) = 0.11

Table 25 shows average weighted values for all evaluation parameters in ALT, ME, NW, and NWS. An average weighted coefficient was then calculated for all parameters. The final weighted average coefficients are as follows:

- ALT: 0.64.
- ME: 0.11.
- NW: 0.12.
- NWS: 0.13.

Table 25. Weighted average coefficients

Parameter	ALT	ME	NW	NWS
Creepage/cycle	0.812	0.090	0.018	0.081
Coating defects/cycle	0.700	0.074	0.120	0.106
Color reduction/cycle	0.541	0.124	0.167	0.168
Gloss reduction/cycle	0.532	0.150	0.167	0.151
Average	0.646	0.110	0.118	0.126

The above weights can be assigned as coefficients to calculate the exposure condition weighted parameter values. For instance, gloss reduction of ASP in ALT at the end of testing is 27.6, which can be multiplied by 0.64 to obtain a weighted gloss reduction of 17.7. Table 26 shows the exposure condition weighted creepage, gloss, color reduction, and values in ALT, ME, NW, and NWS.

Table 26. Weighted coating defects, color, gloss, adhesion, and creepage values

Coating System	Rust Creepage at the Scribe (mm)	Weighted Rust Creepage	Number of Coating Defects	Weighted Coating Defects	Color Reduction, ΔE	Weighted Color Reduction	Gloss Reduction Percent	Weighted Gloss Reduction	Adhesion Strength Variation	Weighted Adhesion Strength Variation
ALT										
Three-coat	5.3	3.39	1	0.64	1.2	0.77	50.9	32.58	0.117	0.07
Two-coat	4.7	3.01	100	64.00	1.4	0.90	60.3	38.59	0.089	0.06
ASP	6.8	4.35	100	64.00	1.4	0.90	27.6	17.66	0.039	0.02
EM	6.5	4.16	100	64.00	8.6	5.50	99	63.36	0.090	0.06
HRCSA	0.7	0.45	2	1.28	6.3	4.03	66.7	42.69	0.011	0.01
HBAC	9.3	5.95	7	4.48	10.9	6.98	79.5	50.88	0.062	0.04
WBEP	15.9	10.18	5	3.20	4.7	3.01	77.8	49.79	0.102	0.07
SLX	21.9	14.02	32	20.48	3.1	1.98	18.5	11.84	0.108	0.07
UM	35.6	22.78	100	64.00	3.7	2.37	23.8	15.23	0.164	0.10
ME										
Three-coat	0	0.00	1	0.11	1	0.11	28.9	3.18	0.158	0.02
Two-coat	1.6	0.18	0	0.00	3.5	0.39	91.5	10.07	0.041	0.00
ASP	1.8	0.20	0	0.00	1.6	0.18	52.7	5.80	0.032	0.00
EM	0.9	0.10	35	3.85	9.6	1.06	97.7	10.75	0.122	0.01

Table 26. (Continued)

Coating System	Rust Creepage at the Scribe (mm)	Weighted Rust Creepage	Number of Coating Defects	Weighted Coating Defects	Color Reduction, ΔE	Weighted Color Reduction	Gloss Reduction Percent	Weighted Gloss Reduction	Adhesion Strength Variation	Weighted Adhesion Strength Variation
HRCSA	1	0.11	0	0.00	9.8	1.08	30.6	3.37	0.011	0.00
HBAC	1.3	0.14	0	0.00	2.2	0.24	29.2	3.21	0.061	0.01
WBEP	1.1	0.12	13	1.43	1.9	0.21	66.9	7.36	0.185	0.02
SLX	30.5	3.36	11	1.21	0.4	0.04	32.8	3.61	0.106	0.01
UM	5.2	0.57	100	11.00	0.4	0.04	4.3	0.47	0.183	0.02
NW										
Three-coat	0	0.00	1	0.12	1	0.12	29.5	3.54	0.157	0.02
Two-coat	1.6	0.19	1	0.12	0.5	0.06	39	4.68	0.061	0.01
ASP	0	0.00	4	0.48	0.3	0.04	10.1	1.21	0.061	0.01
EM	0.6	0.07	12	1.44	14.4	1.73	96.9	11.63	0.144	0.02
HRCSA	0.7	0.08	1	0.12	6.3	0.76	81.9	9.83	0.013	0.00
HBAC	0	0.00	1	0.12	3.3	0.40	24.4	2.93	0.079	0.01
WBEP	0.6	0.07	39	4.68	1.5	0.18	59.3	7.12	0.161	0.02
SLX	2.2	0.26	68	8.16	0.8	0.10	20.6	2.47	0.163	0.02
UM	0.7	0.08	67	8.04	0.2	0.02	1.5	0.18	0.200	0.02
NWS										
Three-coat	0.5	0.07	0	0.00	1	0.13	14.2	1.85	0.188	0.02
Two-coat	1.5	0.20	0	0.00	0.3	0.04	34.5	4.49	0.052	0.01
ASP	0	0.00	0	0.00	0.4	0.05	15	1.95	0.051	0.01
EM	1.6	0.21	12	1.56	15.3	1.99	97.3	12.65	0.134	0.02
HRCSA	0.7	0.09	0	0.00	8.2	1.07	74.1	9.63	0.016	0.00
HBAC	3.7	048	0	0.00	3.3	0.43	16.5	2.15	0.083	0.01
WBEP	2.3	0.30	100	13.00	1.7	0.22	63.8	8.29	0.176	0.02
SLX	12.5	1.63	28	3.64	0.4	0.05	12.4	1.61	0.163	0.02
UM	6.6	0.86	30	3.90	0.2	0.03	0.5	0.07	0.221	0.03

1 inch = 25.4 mm.

Weight of Performance Parameters

Rust creepage, coating defects development, gloss reduction, color variation, and adhesion used for ranking were assigned weight coefficients. The breakdown of weights assignment was based on previous knowledge of performance parameters and their overall impact and significance in evaluating a coating system. GFP was not included due to performance data not being available for outdoor exposure testing.

The performance parameters were assigned the following weights:

- Rust creepage: 0.35.
- Holidays: 0.25.
- Adhesion: 0.10.
- Color reduction: 0.15.
- Gloss reduction: 0.15.

Numerical weighted values of all performance parameters from table 26 for each coating system were averaged to obtain the exposure condition and performance parameter weighted values. Table 27 shows the calculation of final average values for the three-coat system as an example. The resultant values are displayed in row 1 of table 28 excluding the final column. Similar calculations for the remaining coating systems are shown in the rest of table 28.

Weights assigned above were used to calculate the final performance parameter and exposure condition weighted average as follows:

$$\text{Final Average} = \frac{(0.35*\text{Creepage})+(0.25*\text{holidays})+(0.1*\text{Adhesion})+(0.15*\text{color})+(0.15*\text{Gloss})}{5} \text{X}100 \quad (3)$$

Final average or overall parameter was then calculated for ALT, ME, NW, and NWS. Exposure condition and performance parameter weighted values from row 1 were input into the equation as final average to obtain the "Overall Parameter" column in table 28.

Table 27. Average performance parameter calculation for the three-coat system

Performance Parameter	Average Value
Rust creepage	(3.392+0+0+0.065)/4 = 0.864
Coating defects	(0.64+0.11+0.12+0)/4 = 0.218
Color reduction	(0.768+0.11+0.12+0.13)/4 = 0.282
Gloss reduction	(32.576+3.179+3.540+1.846)/4 = 10.285
Adhesion strength	(0.075+0.017+0.019+0.024)/4 = 0.034

Performance Rank

The far right column in table 28 was used to rank the coating systems, and the comprehensive ranking is shown in table 29. Based on the final average values, the coating systems were ranked 1 (best) through 9 (worst). For instance, a coating system with the lowest average was assigned a rank of 1. This is a true characteristic of a coating system since a low final average value indicates good performance.

Table 28. Weighted performance parameters

Coating System	Rust Creepage	Coating Defects	Color Reduction	Gloss Reduction	Adhesion Strength	Overall Parameter
Three-coat	0.864	0.218	0.282	10.285	0.034	38.91
Two-coat	0.893	16.030	0.345	14.456	0.019	130.84
ASP	1.138	16.120	0.290	6.676	0.011	109.42
EM	1.135	17.713	2.569	24.596	0.026	178.05
HRCSA	0.183	0.350	1.733	16.379	0.003	57.37
HBAC	1.644	1.150	2.011	14.791	0.017	67.70
WBEP	2.667	5.578	0.905	18.140	0.032	103.75
SLX	4.815	8.373	0.544	4.883	0.030	91.91
UM	6.075	21.735	0.616	3.988	0.044	165.09

Table 29. Comprehensive rank of one-coat and control systems

Coating System	Rank	Final Average
Three-coat	1	38.91
HRCSA	2	57.37
HBAC	3	67.70
SLX	4	91.91
WBEP	5	103.75
ASP	6	109.42
Two-coat	7	130.84
UM	8	165.09
EM	9	178.05

Evaluation of Coating Systems Based on Ranking

The overall ranking resulted in the three-coat system with the best performance rating followed by HRCSA, HBAC, and WBEP. While positive performance was expected for the three-coat system, the two-coat system was ranked 7th out of all of the coating systems. This reduced ranking can be attributed to the fact that the two-coat system developed an excessive number of coating defects and surface cracking in ALT. The weight coefficient of ALT is the highest among all exposure conditions, which has contributed to the lowered ranking of the two-coat system. Another important reason why the two-coat system may have had a lower comprehensive ranking is due to moderate gloss reduction in ALT and significant gloss reduction in ME, NW, and NWS. Also, it had significantly high rust creepage in ME and NW and moderate rust creepage in NWS. Note that the above described behavior and ranking for the three-coat and the two-coat control systems are based on the evaluation of a particular type of three-coat and two-coat system. The performance and ranking may change if a different type of three-coat or two-coat system were to be evaluated.

Although UM had good color and gloss retention properties in ALT, ME, NW, and NWS, it had high rust creepage and developed blisters, so it was removed from the study after 4,320 h of laboratory testing. EM developed moderate rust creepage and coating defects in outdoor exposures but had very low gloss and color retention properties in all testing conditions. This behavior resulted in low ranking of UM and EM at 8 and 9, respectively.

HBAC and WBEP both demonstrated moderate rust creepage in all exposure conditions except NWS, where WBEP developed high coating defects. Also, WBEP demonstrated higher rusting and blistering in comparison to HBAC. Both of these coating systems showed low color and gloss retention properties, placing them at the lower end of the spectrum. HBAC had a ranking of 3 followed by WBEP at 5.

SLX and ASP were removed from the study after 4,320 h of laboratory testing due to severe blistering and creepage. Although SLX and ASP had moderate to good color and gloss retention, SLX had severe rust creepage in ALT, ME, and NWS with relatively higher rust creepage in NW. ASP had many coating defects that developed in ALT in comparison to SLX. This resulted in a ranking of 4 for SLX and 6 for ASP.

The best performing coating systems were the three-coat system followed by HRCSA. Rust creepage at the scribe followed by the development of coating defects carried the highest weight of coefficients in calculating the final average. Both coating systems had very low rust creepage and little coating defects development, although HRCSA had very low color and gloss retention properties. The three-coat system had higher color and gloss retention properties in comparison to HRCSA, so it had a ranking of 1, followed by HRCSA at 2.

SECTION 4. CONCLUSION

- Although some of the one-coat systems demonstrated promising performance, none of them performed as well as the three-coat system in ALT and the outdoor exposure conditions.
- HRCSA performed well in both ALT and outdoor exposures. While HRCSA is limited by its tenderness for a significant amount of time after application, it presents an interesting alternative for maintenance applications on existing structures.
- Several of the one-coat systems showed encouraging performance in ALT and outdoor exposure conditions in terms of surface failures and rust creepage. GFP and the HBAC were among the better performing candidates.
- Comprehensive performance evaluation showed that the three-coat control was the best performing system, followed by HRCSA, HBAC, and WBEP.
- The two-coat control developed a large number of coating defects in ALT and had significant gloss reduction and rust creepage in outdoor exposure conditions, resulting in a low overall ranking.
- Regression analysis was used to determine the functional relationship between pairs of performance parameters in a specific exposure condition and exposure conditions for a specific performance parameter.
- Color correlated with gloss in all exposure conditions, and coating defects correlated with adhesion strength variation of unscribed panels in NW.
- NW correlated with NWS for color, gloss, and adhesion strength variations. Similarly, adhesion strength variations of unscribed panels in ME correlated with that of NWS.

REFERENCES

[1] Kogler, B. (2008). "Managing the Infrastructure: The Role of Cost Knowledge," *Journal of Protective Coatings and Linings*, 20–31.

[2] Kline, E. S. (2008). "Steel Bridge: Corrosion Protection for 100 Years," *Journal of Protective Coatings and Linings*, 22–31.

[3] Federal Highway Administration. (1995). *Issues Impacting Bridge Painting*, Report No. FHWA-RD-94-098, Federal Highway Administration, Washington, DC.

[4] Chong, S-L. & Yao, Y. (2006). "Are Two Coat as Effective as Three?," *Public Roads*, *70*(2).

[5] Kline, E.S. and Corbett, W.D. (2007). "Completion of Phase I of the Single-Coat Research," *Journal of Protective Coatings and Linings*, 17–18.

[6] ASTM B0117-09. (2010). "Practice for Operating Salt Spray (Fog) Apparatus," *Annual Book of ASTM Standards*, Volume 03.02, ASTM International, West Conshohocken, PA.

[7] ASTM D5894-05. (2010). "Standard Practice for Cyclic Salt Fog/UV Exposure of Painted Metal (Alternating Exposures in a Fog/Dry Cabinet and a UV/Condensation Cabinet)," *Annual Book of ASTM Standards*, Volume 06.01, ASTM International, West Conshohocken, PA.

[8] Richard, S.H. (1995). "Research News: New England DOTs Develop Program for Selecting Coating Systems," *Journal of Protective Coatings and Linings*, 17–39.

[9] Chong, S-L., Jacoby, M., Boone, J., and Lum, H. (1995). *Comparison of Laboratory Testing Methods for Bridge Coatings*, Report No. FHWA-RD-94-112, Federal Highway Administration, Washington, DC.

[10] Ault, J.P & Farschon, C.L. (2009). "20-Year Performance of Bridge & Maintenance Systems," *Journal of Protective Coatings and Linings*, 16–32.

[11] Chong, S-L & Yao, Y. (2003). *Laboratory Evaluation of Waterborne Coatings on Steel*, Report No. FHWA-RD-03-032, Federal Highway Administration, Washington, DC.

[12] Angeloff, C., Squiller, E. P. & Best, K. E. (2002). "Two-Component Aliphatic Polyurea Coatings for High Productivity Applications," *Journal of Protective Coatings and Linings*, 42–46.

[13] Chong, S-L & Yao, Y. (2007). "Selecting Overcoats for Bridges," *Public Roads, 71*(2).

[14] Mowrer, N. R. (2005). *Polysiloxane Coating Innovations*, PPG Protective & Marine Coatings, Pittsburgh, PA. Obtained from: http://ppgamercoatus.ppgpmc.com/techcenter/resources.cfm?resource=2.Site last accessed May 12, 2011.

[15] SSPC-SP-10. (2007). *The Society for Protective Coatings Surface Preparation Specification Near-White Metal Blast Cleaning*, The Society for Protective Coatings, Pittsburgh, PA.

[16] ASTM D1654-08. (2010). "Standard Test Method for Evaluation of Painted or Coated Specimens Subjected to Corrosive Environments," *Annual Book of ASTM Standards*, Volume *06.01*, ASTM International, West Conshohocken, PA.

[17] ASTM G85-09. (2010). "Standard Practice for Modified Salt Spray (Fog) Testing," *Annual Book of ASTM Standards*, Volume *03*.02, ASTM International, West Conshohocken, PA.

[18] Ault, P., Ellor, J., Repp, J. & Shaw, B. (2000). *Characterization of the Environment*, Report No. FHWA-RD-00-030, Federal Highway Administration, Washington, DC.

[19] ASTM D2369–10. (2010). "Standard Test Method for Volatile Content of Coatings," *Annual Book of ASTM Standards*, Volume *06.01*, ASTM International, West Conshohocken, PA.

[20] ASTM D2371–85. (2010). "Standard Test Method for Pigment Content of Solvent-Reducible Paints," *Annual Book of ASTM Standards*, Volume *06.01*, ASTM International, West Conshohocken, PA.

[21] ASTM D3723-05e1. (2010). "Standard Test Method for Pigment Content of Water-Emulsion Paints by Low-Temperature Ashing," *Annual Book of ASTM Standards*, Volume *06.01*, ASTM International, West Conshohocken, PA.

[22] ASTM D4400–99. (2010). "Standard Test Method for Sag Resistance of Paints Using a Multinotch Applicator," *Annual Book of ASTM Standards*, Volume *06.02*, ASTM International, West Conshohocken, PA.

[23] ASTM D1640-03. (2010). "Standard Test Methods for Drying, Curing, or Film Formation of Organic Coatings at Room Temperature," *Annual Book of ASTM Standards*, Volume *06.01*, ASTM International, West Conshohocken, PA.

[24] SSPC-SP-2. (1982). *The Society for Protective Coatings Surface Preparation Specification No. 2 Hand Tool Cleaning*, The Society for Protective Coatings, Pittsburgh, PA.

[25] Nadal, M. E. (2001). "NIST Reference Goniophotometer for Specular Gloss Measurements," *Journal of Coatings Technology*, *73*(917), 73–80.

[26] ASTM D523-08. (2010). "Standard Test Method for Specular Gloss," *Annual Book of ASTM Standards*, Volume *06.01*, ASTM International, West Conshohocken, PA.

[27] ASTM D2244-09A. (2010). "Standard Practice for Calculation of Color Tolerances and Color Differences from Instrumentally Measured Color Coordinates," *Annual Book of ASTM Standards*, Volume *06.01*, ASTM International, West Conshohocken, PA.

[28] ASTM D3363-05. (2010). "Standard Test Method for Film Hardness by Pencil Test," *Annual Book of ASTM Standards*, Volume *06.01*, ASTM International, West Conshohocken, PA.

[29] ASTM D4541-09. (2010). "Standard Test Method for Pull-Off Strength of Coatings Using Portable Adhesion Testers," *Annual Book of ASTM Standards*, Volume *06.02*, ASTM International, West Conshohocken, PA.

[30] ASTM D5162-08. (2010). "Standard Practice for Discontinuity (Holiday) Testing of Nonconductive Protective Coating on Metallic Substrates," *Annual Book of ASTM Standards*, Volume *06.02*, ASTM International, West Conshohocken, PA.

[31] ASTM D714–02. (2010). "Standard Test Method for Evaluating Degree of Blistering of Paints," *Annual Book of ASTM Standards*, Volume *06.01*, ASTM International, West Conshohocken, PA.

[32] ASTM D610–08. (2010). "Standard Practice for Evaluating Degree of Rusting on Painted Steel Surfaces," *Annual Book of ASTM Standards*, Volume *06.02*, ASTM International, West Conshohocken, PA.

[33] ASTM D7087-05A. (2010). "Standard Test Method for An Imaging Technique to Measure Rust Creepage at Scribe on Coated Test Panels Subjected to Corrosive Environments," *Annual Book of ASTM Standards*, Volume *06.01*, ASTM International, West Conshohocken, PA.

[34] Hare, C.H. (1994). *Protective Coatings: Fundamentals of Chemistry and Compositions*, SSPC Publication, Pittsburgh, PA.

[35] Hare, C.H. (2001). *Paint Film Degradation: Mechanisms & Control*, SSPC Publication, Pittsburgh, PA.

[36] Brockman, W. (1981). *Durability of Metal Polymer Bonds in Adhesion Aspects of Polymeric Coatings*, Proceedings of the Symposium on Adhesion Aspects of Polymeric Coatings, Minneapolis, MN.

INDEX

A

B

C

D

E

F

G

H

I

J

L

M

N

O

P

Q

R